CHEMICAL ENGINEERING
PRACTICE EXAM SET

Second Edition

Randall N. Robinson, P.E.

PROFESSIONAL PUBLICATIONS, INC.
Belmont, CA 94002

In the ENGINEERING REVIEW MANUAL SERIES

Engineer-In-Training Review Manual
  Engineering Fundamentals Quick Reference Cards
  Mini-Exams for the E-I-T Exam
  1001 Solved Engineering Fundamentals Problems
  E-I-T Review: A Study Guide
Civil Engineering Reference Manual
  Civil Engineering Quick Reference Cards
  Civil Engineering Sample Examination
  Civil Engineering Review Course on Cassettes
  Seismic Design for the Civil P.E. Exam
  Timber Design for the Civil P.E. Exam
Structural Engineering Practice Problem Manual
Mechanical Engineering Review Manual
  Mechanical Engineering Quick Reference Cards
  Mechanical Engineering Sample Examination
  Mechanical Engineering Review Course on Cassettes
  101 Solved Mechanical Engineering Problems
  Consolidated Gas Dynamics
Electrical Engineering Review Manual
Chemical Engineering Reference Manual
  Chemical Engineering Practice Exam Set
Land Surveyor Reference Manual
Metallurgical Engineering Practice Problem Manual
Petroleum Engineering Practice Problem Manual
Expanded Interest Tables
Engineering Law, Design Liability, and Professional Ethics
Engineering Unit Conversions

In the ENGINEERING CAREER ADVANCEMENT SERIES

How to Become a Professional Engineer
The Expert Witness Handbook—A Guide for Engineers
Getting Started as a Consulting Engineer
Intellectual Property Protection—A Guide for Engineers
E-I-T/P.E. Course Coordinator's Handbook

Distributed by: Professional Publications, Inc.
1250 Fifth Avenue
Department 77
Belmont, CA 94002
(415) 593-9119

CHEMICAL ENGINEERING PRACTICE EXAM SET
Second Edition

Printed in the United States of America

ISBN: 0-932276-93-8

Professional Publications, Inc.
1250 Fifth Avenue, Belmont, CA 94002

Current printing of this edition (last number): 6 5 4 3 2 1

TABLE OF CONTENTS

PREFACE

"It is not always by plugging away at a difficulty and sticking at it that one overcomes it; but, rather, often by working on the one next to it. Certain people and certain things require to be approached on an angle."[1]

Within this quotation lies the method for being successful when taking the Professional Engineering Exam. To be successful, the student must choose four problems from a list of ten to solve. This is the key: too often the student chooses the wrong problems. Those problems should be the easiest ones for the student to solve.

Collected in this second edition are the exams that I give to my students each year near the end of the course, usually about two weeks before the actual exam. These exams represent the degree of difficulty and topics which you will probably encounter in the exam.

There are certain characteristics you should be aware of when reviewing these sample examinations. First, you may notice that one or two problems are repeated; the real exam also repeats some problems. Other problems are not specifically covered in the *Chemical Engineering Reference Manual*; these occur infrequently. Some problems cannot be solved in one hour (problem 5-15 is a good example); the teaching point here is to make you aware that problems like this do exist on the exam. Avoid them at all costs (reread the opening quotation). Finally, don't be surprised to find one or two easy problems on the exam; they do exist, so don't look for hidden meanings. This final characteristic can be illustrated by another of my favorite quotations, *"Attempt easy tasks as if they were difficult, and difficult as if they were easy: in the one case that confidence may not fall asleep, in the other that it may not be dismayed."*[2]

Good overall reading of the exam at first helps give the student a plan of attack. Careful and liberal inspection for overall review never is a problem. This is the answer.

I suggest you try every exam in this supplement under exam conditions. At the very least, become familiar with all the problems and their solutions.

Randall N. Robinson, P.E.
San Jose, CA
September, 1988

---

[1] André Gide, *Journals*, October 26, 1924

[2] Baltasar Gracian, *The Art of Worldly Wisdom*, 1647

Chemical Engineering Professional License Sample Exam #1 A

1-1 A city engineer must determine the capitalized cost of perpetual service associated with constructing and maintaining a new water storage facility. The facility will be constructed in two stages. The first water tank will have an initial cost of $24,000. The second tank will be built 10 years later at a cost of $40,000. When the second tank is installed, a $7000 pumping station must also be provided. The pumps have a life of seven years. Maintenance costs will run $800 during the first year of pump life and will increase $100 each year until the end of the service life for any given pump. Annual costs will return to $800 after seven years when the pumps are replaced. An interest rate of 10% is specified for all analyses.

1-2 A colloidal suspension in water is cooled from 150 °F to 50 °F. The stream is allowed to cool naturally in a holding tank. Another process requires a benzene feed stream to be heated from 60 °F to 110 °F. The benzene is heated with steam in a 1-2 heat exchanger with a 500 ft^2 heat exchange area. Can the heat exchanger be used to recover heat in the colloid by heating benzene? The two streams are 10,000 lbm/day colloid and 25,000 lbm/day benzene. Assume the overall heat transfer coefficient based on outside area, U_o = 100 BTU/hr-ft^2-°F, specific heat of the colloid = 1.0 BTU/lbm-°F, and specific heat of benzene = 0.44 BTU/lbm-°F.

1-3 A batch reactor is used to react acetic acid (HAc) and butanol to form the acetate. The following data are given:

component	charge	molecular weight
HAc	106 lbm	60
butanol	650 lbm	74
total	756 lbm	

$$\rho = 0.75 \text{ gm/ml}$$

reaction: second-order, irreversible
reaction time: 53 minutes
mole ratio: butanol/HAc = 4.98

K = 17.4 ml/gmole-min at 100 °C
K = 30.0 ml/gmole-min at 120 °C

Which results in the least reaction time: (a) increase temperature to 120 °C or (b) change mole ratio to 4.0, keeping the gmoles HAc constant?

1-4 A pump supplies water at 80 °F to a scrubbing tower. The water is pumped through a 2″ schedule-40 welded steel pipe with 220 feet equivalent length. The scrubber inlet is 60 feet above the level of the water in the tank. Determine the flow coefficient, C_v, of the control valve to allow throttling of the flow to deliver 50 to 140 gpm.

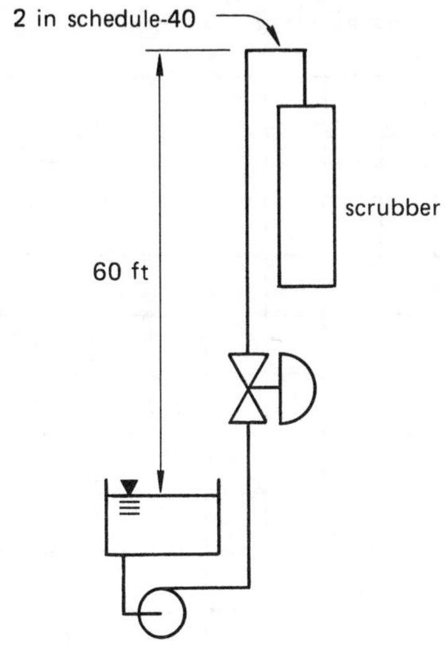

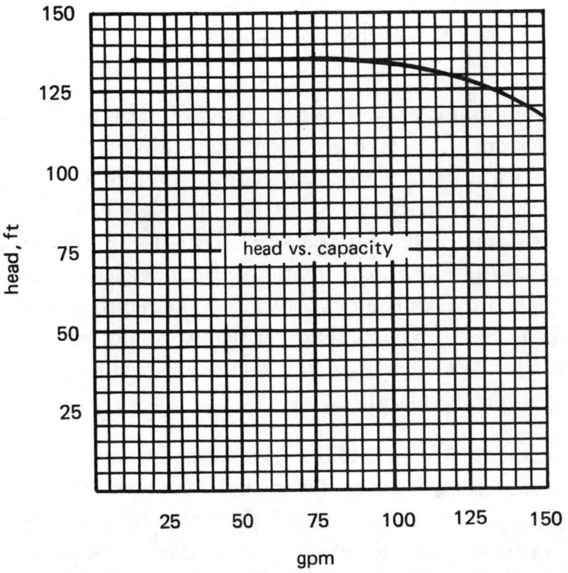

1-5 The oxidation of nitrogen to form NO has an equilibrium constant of 0.002 at 2000 °F. The reaction is:

$$\tfrac{1}{2}N_2 + \tfrac{1}{2}O_2 \longrightarrow NO$$

What will be the mole% of NO if air is heated to 2000 °F? Use an air composition of 21% O_2 and 79% N_2.

1-6 A packed tower separates benzene and toluene. The overhead product is 92.4 weight% benzene and contains 90% of the benzene in the feed. The column operates at 1.3 times the minimum reflux ratio. The feed is 60% benzene and enters as liquid at 275 °C. The tower is packed with $\frac{1}{4}''$ stainless steel Raschig rings. Use the data given to determine the height of the packing required.

boiling point of feed = 270 °C

λ = 370 J/g for benzene and toluene

HETP = 0.2 meters

c_p = 1.3 J/g-°C for benzene and toluene

packing efficiency = 60%

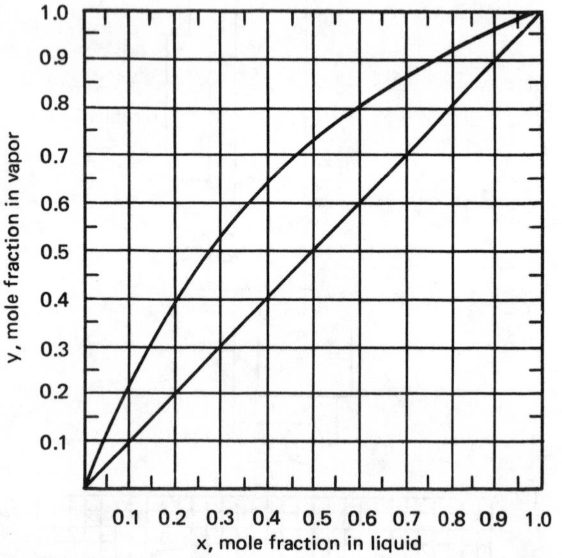

1-7 A company wishes to produce 2000 lbm/hr of 65% NaOH from a 5% feed solution. A single-effect evaporator with a 2500 ft^2 heat exchanger area is available, with steam available at 5, 40, or 100 psig. The overall heat transfer coefficient is 300 BTU/hr-ft^2-°F. The feed enters at 70 °F, and the evaporator condenser operates at 1 psia. (a) Can the evaporator do the required work? (b) If it does, what steam pressure would you use? Why? (c) Which parameter would you use to control the system? Explain.

1-8 Air at 230 °F, 0.006 lbm H_2O/lbm dry air is fed to a countercurrent dryer to dry granular material that contains 40% moisture and is fed at 1000 lbm/hr at 60 °F. Air exits at 100 °F, and the granular material exits at 95 °F and 5% moisture. The latent heat of water vapor at 68 °F is 1052 BTU/lbm. What air rate is required, and what is the outlet humidity? The following data are given:

c_p water vapor = 0.48

c_p air = 0.231

c_p dry material = 0.20

1-9 For the simple multistage extraction process in the figure, find the second-stage raffinate composition and the percent of component C extracted in the first stage. Use the equilibrium diagram provided.

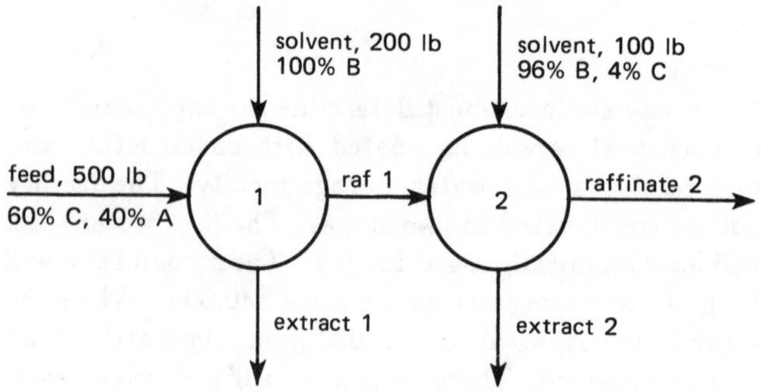

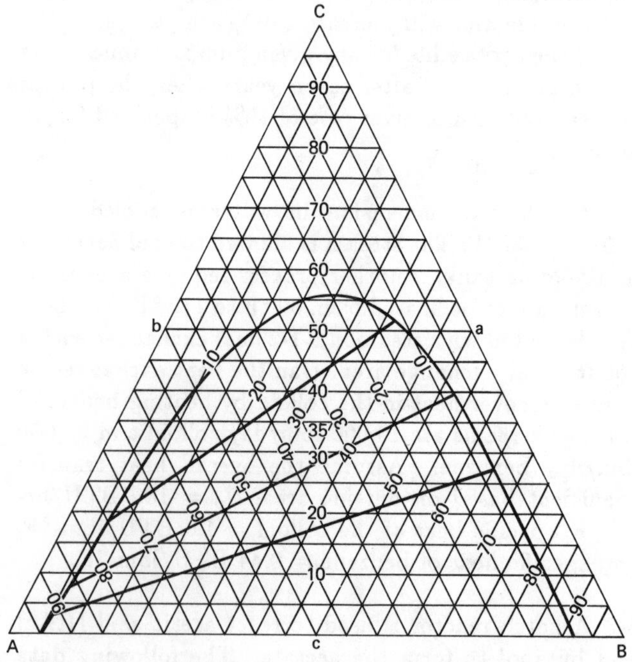

1-10 A continuously stirred tank reactor (CSTR) is used in a laboratory experiment to study the alkylation of ethyl acetate (EtAc) with NaOH. The following data are collected:

reactor volume = 602 ml

inlet conditions:

NaOH = 1.6 liter/hr; [OH] = 0.00587$\underline{M}$

EtAc = 1.2 liter/hr; [Ac] = 0.0389 $\underline{M}$

outlet conditions:

[OH] = 0.001094 $\underline{M}$ at 25 °C

[OH] = 0.0009243 $\underline{M}$ at 40 °C

The reaction is first-order with respect to hydroxide and first-order with respect to acetate. Develop an expression for the reaction rate constant as a function of temperature.

Chemical Engineering
Professional License Sample Exam
#1 B

1-11 The ideal gas (zero pressure) heat capacity of chlorine is given by the equation:

$$c_p = 6.8214 + 0.0057095\,T - 5.107 \times 10^{-6}T^2 + 1.547 \times 10^{-9}T^3$$

where specific heat at constant pressure, c_p, is given in cal/gmole-K and T is in K. What is the heat capacity at 140 °F and 112 psia?

1-12 A distillation column separates components A and B from a binary mixture. The column products are: 0.3 mole/hr overhead at 95 mole% A, 0.5 mole/hr bottoms, and 0.2 mole/hr side stream at 75 mole% A. Feed enters at 47 mole% A liquid at its boiling point. A reflux ratio of 2 is used at the top of the column. Additional reflux will be provided by removing 0.4 mole/hr vapor as the side stream, condensing it, and returning it to the column as 0.2 mole/hr liquid. The volatility ratio is 2.4. (a) What is the number of theoretical plates in the column? (b) What is the location of the feed plate? (c) What is the location of the side stream?

1-13 A 10,000 gallon vertical cylindrical tank will contain NaOH with a specific gravity of 1.525. The tank must be a minimum of $\frac{3}{16}''$ thick for structural strength. What is the optimum tank size and cost?

cost parameters

top: $4/ft²

bottom: $9/ft² (including pad)

wall: $21 t/ft² where $t = \dfrac{pd}{24{,}000}$

p = hydrostatic pressure, psi
d = tank diameter, inches
t = wall thickness, inches

1-14 A finned tube heat exchanger in a warehouse heats air with process steam at a total pressure of 1 atm. A $1\frac{1}{2}''$ schedule-40 steel pipe is used with enough fins to give 3 ft² of area per foot of length (2.5 ft² for fins, 0.5 ft² for the pipe). The steam quality is 0.95 and the condensate exits at 200 °F. (a) How much steam is needed if 60,000 BTU/hr is required to maintain the room at 70 °F? (b) What length pipe is needed if the air temperature rise in the heat exchanger is 30 °F? Use the following data:

$$\begin{aligned} h_{\text{air}} &= 5 \text{ BTU/hr-ft}^2\text{-°F} \\ h_{\text{steam}} &= 200 \text{ BTU/hr-ft}^2\text{-°F} \\ \text{fin efficiency} &= 80\% \\ (c_p)_{\text{air}} &= 0.25 \text{ BTU/lbm-°F} \end{aligned}$$

1-15 In a single continuously stirred tank reactor (CSTR), the following irreversible second-order reaction takes place with 60% conversion:

$$\text{A} + \text{B} \longrightarrow \text{R} + \text{S}$$

The following data apply:

$$\begin{aligned} \text{volume of reactor} &= 2 \text{ ft}^3 \\ \text{flow} &= 8 \text{ ft}^3\text{/min} \\ C_{Ao} &= 2 \text{ lbmole/ft}^3 \\ C_{Bo} &= 3 \text{ lbmole/ft}^3 \end{aligned}$$

Assuming constant density and no temperature change in the reactor, (a) what will be the overall conversion of A, X_A, if a second identical reactor is added in series? (b) What will be the composition of the effluent from the second reactor?

1-16 A pond feeds a sprinkler system by gravity with a required minimum flow of 1600 gpm. After an 8″ schedule-80 line was installed (total length of 3000 feet), the flow was measured as 1765 gpm. A year later the sprinkler system was extended from the same pond. At 900 feet from the discharge point of the original 8″ line, a new line was installed with a minimum flow requirement of 300 gpm. Given a water temperature of 60 °F, will the new system be capable of supplying the required capacity? The Darcy friction factor used for this system was calculated using the following formula:

$$f = 0.0035 + 0.027(N_{\text{Re}})^{-0.42}$$

1-17 For the reaction given at 800 K with stoichiometric reactants $CO + 3H_2 \longrightarrow CH_4 + H_2O$, calculate the (a) equilibrium conversion of methane and (b) the heat added or removed at equilibrium conversion. Use the data given:

	T, K	$-(G_o^o - H_o^o)/T$ kcal/gmole K	$(H_f^o)_o$ kcal/gmole	$(H_T^o - H_o^o)$ kcal/gmole
$CO_{(g)}$	800	47.254	−27.02	5.7000
$H_{2(g)}$	800	31.186	0	5.5374
$CH_{4(g)}$	800	45.21	−15.987	8.321
$H_2O_{(g)}$	800	45.128	−57.107	6.6896

1-18 Specify the height of packing and the tower diameter required in an absorption tower to remove 99% of the N_2O_4 in a gas. The packing will be $\frac{1}{2}''$ porcelain Raschig rings. The gas enters at a rate of 1 lbm/sec and contains 1.0 weight% N_2O_4 and 99% wet air at 300 K. The absorbant consists of water entering at 0.2 lbm/sec. The N_2O_4 reacts readily with water. Assume $k_y a = 100$ lbm N_2O_4/hr-ft^3-Δy, and the capacity of the column at the flood point is 1000 lbm gas/hr-ft^2. The units of y are lbm N_2O_4/lbm air.

1-19 Calculate the endpoints and three intermediate points on the vapor-liquid equilibrium curve (y vs. x) for binary mixtures of propane and n-butane at a total pressure of 5 atm, assuming ideal behavior.

1-20 A binary mixture of components A and B containing 50 mole% A is distilled in a column containing 9 theoretical trays in addition to a reboiler and a condenser. The feed enters below the bottom tray directly into the reboiler. A reflux ratio of 3 is used. The distillate contains 96 mole% A (the more volatile component) and $\alpha_{AB} = 2.0$. (a) What mole fraction A is in the bottoms? (b) What fraction of the feed will be produced in the bottoms if the column is operated in the proper manner?

Chemical Engineering Professional License Sample Exam #2 A

2-1 Consider a prospective investment in a project with an initial cost of $300,000, annual operating and maintenance costs of $35,000 per year, and an estimated net disposal value of $50,000 at the end of 60 years. Assume an interest rate of 8%. (a) What is the present value of this investment if the planning horizon is 60 years? (b) What is the capitalized cost of perpetual service if the project replacement has the same initial cost, life, salvage value, and operating costs as the original?

2-2 A gas mixture containing air and ammonia is scrubbed with water in a packed column. Gas enters the column at 20 moles/hour; the ammonia mole fraction is 0.005; 20 moles of water are used per hour. The tower has an inside diameter, I.D., of 2 feet, has a height of 4 feet, and operates at 25 °C, 1 atm. Packing is $\frac{1}{2}''$ Raschig rings. For this packing and the ammonia-air system at 1 atm total pressure, $H_G = 5.31\,G^{0.1}\,L^{-0.39}$, where H_G is in feet and G and L are in lbm/hr-ft^2. The equilibrium relationship is expressed by $p = X$, where p is the ammonia pressure in atm and X is the mole fraction. Neglect water vapor and assume the gas film controls. What percent ammonia recovery can be expected using (a) countercurrent flow and (b) parallel flow?

2-3 A factory has 10,000 lbm/hr of hot water (350 °F) that would normally be discarded. Heat from this hot stream is recovered by using it to preheat 10,000 lbm/hr of distilled water from 70 °F. This distilled water will ultimately be heated to 400 °F. What is the approximate distilled water outlet temperature that could reasonably be obtained and the heat that could reasonably be recovered for (a) a co-current exchanger, (b) a counterflow exchanger, and (c) a 1-2 heat exchanger with the distilled water in the tubes?

2-4 A 10% (± 0.2%) by weight NaOH solution is used for a caustic washing process that utilizes a 10,000 gallon surge tank to smooth variations in flow rate and concentration. The process requires a constant flow rate of 1500 gal/hr of solution. Just before a plant upset, the tank contains 8000 gallons of 10% NaOH with an inlet flow of 1000 gal/hr of 10.5% NaOH to the tank. Due to the upset, the inlet concentration drops to 5% NaOH instantly, although all flows remain constant. An alarm sounds when the outlet concentration from the surge tank drops below 9.8% (or rises to 10.2%). The tank is well mixed, and the density of all solutions is 8.8 lbm/gal. All concentrations are on a weight basis. (a) How long will it take before the alarm sounds? (b) If the alarm does not operate, how long will it take before the tank is empty?

2-5 Air conditioning equipment for a paper testing laboratory building will be designed to operate as shown.

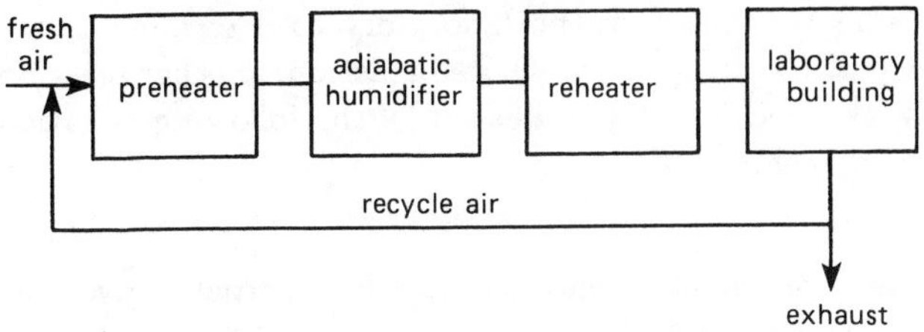

The design conditions are:

air to laboratory building: 1,000,000 ft^3/hr at 72 °F and 50% relative humidity

recycle air: 500,000 ft^3/hr at 70 °F and 50% relative humidity

fresh air: 30 °F and 10% relative humidity

steam: 25 psig

The humidifier must be designed so that the air leaving it is at 80% relative humidity. The following additional data are available:

for the humidifier: $h_y a = 90$ BTU/ft^3-hr-°F
for the preheater: $U = 12$ BTU/ft^2-hr-°F

Calculate (a) the quantity of fresh air required, in cfm, at entrance conditions, (b) the temperature and absolute humidity of the air at the entrance of the preheater, the entrance of the humidifier, and the exit of the humidifier, (c) the water temperature required in the humidifier, and (d) the volume required for the humidifier.

2-6 An atmospheric distillation column, equipped with a total condenser and a kettle reboiler, is operated to separate a mixture containing 50 mole% A (the more volatile component) and 50 mole% B. The overhead stream is 35 moles/100 moles of feed, with a composition of 95 mole% A, and the feed is a vapor-liquid mixture of about 50% quality.

It is necessary to increase the yield of overhead product (decrease the A content of the bottoms) at the same overhead composition. Assume that the column is not limiting in any way. (a) If condenser and reboiler capacity are adequate, prove whether or not the objective can be achieved by increasing the reflux ratio. (b) If reboiler capacity is adequate but condenser capacity prevents any increase in overhead rate, prove whether or not the desired result can be achieved by installing a cooler to cool the feed to saturated liquid.

2-7 A liquid-phase, constant volume reaction, with stoichiometry indicated by the equation:

$$A + B \longrightarrow R$$

is being studied in the laboratory to determine a rate equation. Besides the desired reaction, another reaction with stoichiometry indicated by the following equation also occurs:

$$R + B \longrightarrow S$$

An experimental run made in a batch reactor gives the following results:

time, min	C_A gmoles/liter	C_B gmoles/liter
0	1.40	3.00
10	0.89	2.41
15	0.74	2.23
20	0.64	2.07
25	0.55	1.92
30	0.47	1.80
40	0.37	1.61
50	0.29	1.45
60	0.23	1.32
70	0.19	1.22

It is suspected that the reaction rates can be expressed by the two rate equations:

$$-r_A = \frac{-dC_A}{d\theta} = k_1 C_A C_B$$

$$\text{and} \quad r_S = \frac{dC_S}{d\theta} = k_2 C_R C_B$$

(a) Prove that these data are consistent with the proposed rate of expressions. (b) Determine values of k_1 and k_2.

2-8 A pitot tube is installed in a 3″ schedule-40 pipe to determine water flow. The water temperature is 60 °F. The manometer fluid is mercury, and for the existing steady state conditions, R reads 1.4″. What is the mass flow rate, in lbm/sec, of water through the pipe?

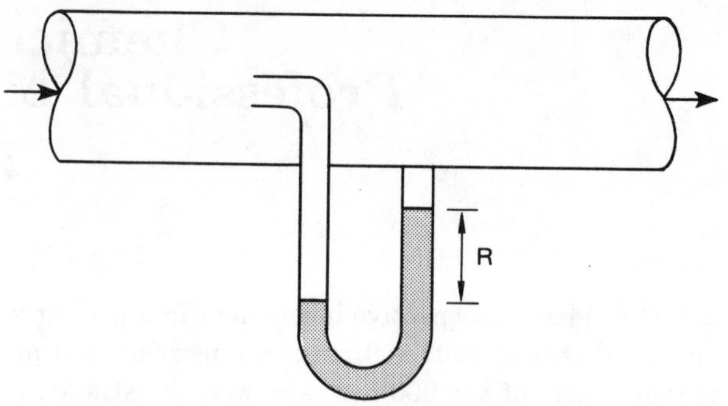

2-9 Design a plant to produce component S by the gas phase decomposition over a solid catalyst. The reaction is:

$$A_{(g)} \rightleftharpoons R_{(g)} + S_{(g)}$$

Bench scale studies using the company's catalyst indicate that the rate of reaction may be approximated by the equation:

$$r_A = \frac{C\left(\frac{p_A - p_R p_S}{K}\right)}{(1 + K_A p_A + K_{RS} p_{RS})^2}$$

r_A = reaction rate, lbmoles A converted/lbm of catalyst/hour

p_A = partial pressure of A, atm

p_R = partial pressure of R, atm

p_S = partial pressure of S, atm

$p_{RS} = \frac{p_R + p_S}{2}$

C = rate constant, lbmoles A/hr-lbm-atm of catalyst

K = overall gas phase equilibrium constant, atm

K_A = adsorption equilibrium constant for A, atm^{-1}

K_{RS} = average effective adsorption equilibrium constant for R and S, atm^{-1}

Ignore possible side reactions to simplify your calculations. The reactor will operate at 0.10 atm total pressure and 1060 °F. Enough heat can be supplied from outside to keep the reactor isothermal. Neglect pressure drop due to flow through the catalyst bed. The feed is 20 lbmoles/hr of pure A at 1060 °F. How much catalyst (lbm) must be used in order to get a conversion of 0.2 lbmole of A/lbmole of A fed?

Under the proposed operating conditions, the following equations apply:

$$C = 0.0696 \text{ lbmoles/hr-lbm-atm}$$

$$K_A = 0.44 \text{ atm}^{-1}$$

$$K_{RS} = 1.524 \text{ atm}^{-1}$$

$$K = 0.201$$

2-10 The flow diagram for a process used to convert *n*-hexane, *n*-C_6H_{14}, to cyclohexane, C_6H_{12}, is shown in the figure. The feed passing through the preheating furnace is pure *n*-hexane entering at 77 °F. In the furnace it is preheated to 300 °F before entering the catalytic reactor, where 50% is converted to cyclohexane by straight dehydrogenation:

$$C_6H_{14} = C_6H_{12} + H_2$$

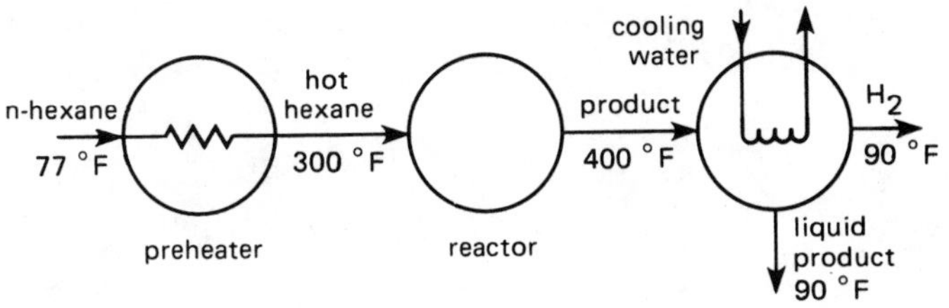

The product leaves the catalytic reactor at 400 °F and is cooled to 90 °F with condensation of *n*-hexane and cyclohexane. The pressure is 1 atm, and there are no heat losses in the preheater or condenser cooler. Pressure drop attributable to flow through the system may be neglected. On the basis of 100 lbm of *n*-hexane feed, (a) how many BTU of heat are required in the preheater and (b) how much heat is evolved or absorbed (specify which) in the reactor? Assume no heat loss by radiation.

- Heat capacities in BTU/lbm-°F are:

n-hexane liquid	0.60
n-hexane vapor	0.47
cyclohexane liquid	0.45
cyclohexane vapor	0.40
hydrogen	3.50

- Normal boiling points are 156 °F for *n*-hexane and 180 °F for cyclohexane.
- Heats of vaporization at the normal boiling point are 6900 cal/gmole for *n*-hexane and 7190 cal/gmole for cyclohexane.
- Heats of combustion at 25 °C (77 °F) and 1 atm are:

	ΔH_c
n-hexane (liquid)	994,600 cal/gmole
cyclohexane (liquid)	937,800 cal/gmole
hydrogen (gas)	68,372 cal/gmole

- Products of combustion are $CO_{2(g)}$ and $H_2O_{(l)}$.

Chemical Engineering Professional License Sample Exam #2 B

2-11 A pond/dam arrangement where the level of the water is maintained at an elevation of 3870 feet has a water intake 10 feet below the pond surface. The intake line is 6″ schedule-40 galvanized iron pipe 350 feet long, leading to a turbine generator unit at an elevation of 3780 feet. The turbine discharge pipe (the same type as the intake pipe) is 2050 feet long and discharges into the air at an elevation of 3750 feet. A control valve in the discharge line is set to permit the flow of 58.0 ft^3/min at a water temperature of 60 °F when the pipe is running full. Turbine efficiency is 86%, and fitting and entrance/exit losses amount to 815 pipe diameters. (a) Determine the total frictional loss per lbm of fluid and the shaft work per lbm of fluid. (b) How many 100 W light bulbs could be lit by this system?

2-12 A 1-2 heat exchanger heats 50 gpm of city water from 60 °F to 140 °F, using 200 gpm of hot water entering the shell at 250 °F. The designer used a dirt factor of 0.0022 and a design dirty overall heat transfer coefficient based on outside area, U_o, of 292. Estimate the outlet temperatures of the two streams, assuming no effect of temperature on U_o, at design inlet conditions when the exchanger is first put into service.

2-13 Air is cooled and dehumidified by countercurrent contact in a packed tower with chilled water. The design conditions are:

inlet air rate:	10,000 lbm/hr of dry air
inlet air conditions:	90 °F dry bulb
	80 °F wet bulb
inlet water rate:	20,000 lbm/hr
inlet water temperature:	60 °F
outlet water temperature:	68 °F

Assume that the air leaves the tower saturated at its exit temperature. (a) What is the exit air temperature and humidity? (b) How much water, in lbm/hr, is removed from the air?

2-14 A reciprocating compressor compresses 100 cfm of dry air from 1 atm and 90 °F to 6 atm. Clearance on the compressor is 0.05, and the isentropic compression efficiency is 82%. Calculate the actual horsepower required to drive the compressor and the piston displacement in cubic feet if the compressor is double-acting and operating at 200 strokes/min.

2-15 A reaction having the following stoichiometry will be run commercially with a large excess of component B.

$$A + B \longrightarrow \text{products}$$

Under these conditions the rate equation is:

$$-r_A = kC_A$$

The design conversion is 95%. The reaction is strongly exothermic and close temperature control is required. It is, therefore, necessary that the reaction be carried out in dilute solutions using one or more mixed flow (CST) reactors. In the range of interest, the capital cost of the reactors is a function of the capacity of the reactor according to the equation:

$$\text{FCI} = (A)V^{0.65}$$

FCI is the fixed capital investment in dollars, A is a constant, and V is the capacity of a single reactor in gallons. (a) How many reactors should be used? (b) How should they be arranged for minimum capital investment?

2-16 An insoluble fibrous material is contaminated with a heavy oil at a 4:1 ratio per lbm oil to lbm fiber. This material will be fed continuously to a countercurrent cascade where a pure wash solvent will contact it. 98% of the oil is to be removed. The solvent is miscible with the oil in all proportions.

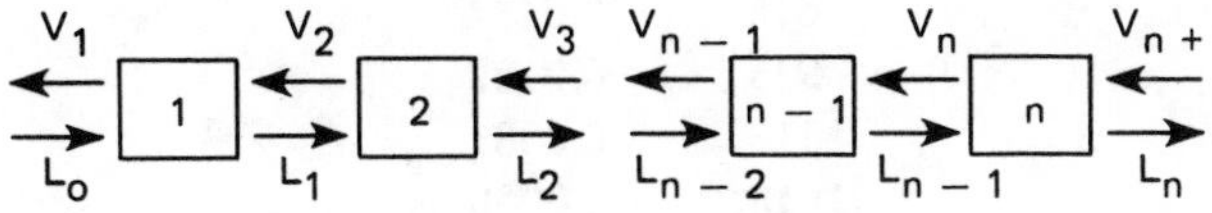

L_o = 100 lbm/hr
fiber = 20 lbm/hr
oil = 80 lbm/hr

The locus of underflow compositions is shown on the tenary diagram. (a) How much insoluble fibrous material will leave each stage in the underflow (raffinate)? (b) How much oil will be present in L_n? (c) How much solvent will be present in L_n? (d) What is the minimum feed rate of fresh solvent that will effect the desired separation of oil from the fibers? (e) How many theoretical stages are necessary to effect the desired separation if a solvent feed rate 1.3 times the minimum is used?

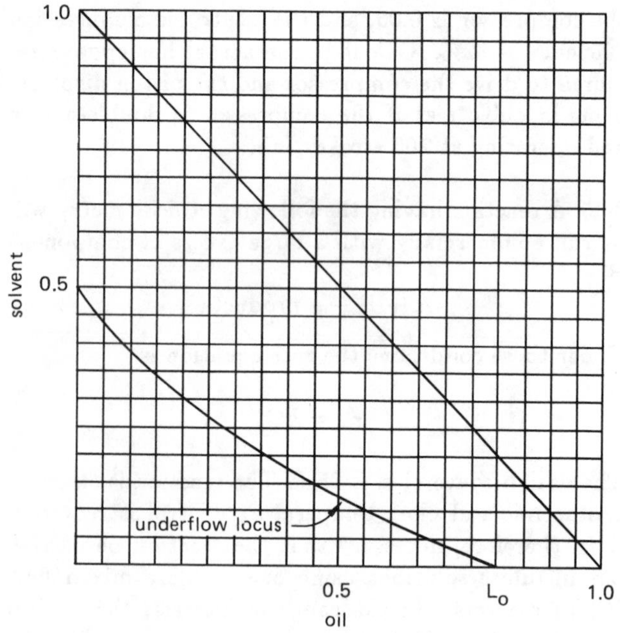

2-17 An experiment is performed to determine the kinetics for the reaction between ethyl acetate (EtAc) and NaOH to form NaAc and ethanol (EtOH) according to the reaction:

$$\text{EtAc} + \text{NaOH} \longrightarrow \text{NaAc} + \text{EtOH}$$

One liter of 0.0912 N NaOH is added to one liter of 0.0852 N EtAc and stirred in a batch reactor at 25 °C. The concentration of the NaOH is monitored with time:

time (sec)	NaOH (gmole/liter)
90	0.0373
180	0.0291
270	0.0231
615	0.0134
915	0.0122
1515	0.0081
1815	0.0063

If the reaction is first-order with respect to both the NaOH and EtAc, determine the value of the reaction rate coefficient. Indicate units in your answer.

2-18 A centrifugal pump pumps 180 °F water at a rate of 200 gpm from a tank vented to the atmosphere. The outlet from the tank is 5 feet above the pump. The pump suction line is an 8 foot length of 4″ schedule-40 pipe. This suction line contains two short radius 90° elbows. What is the net positive suction head that will be available to the pump when the water level is at the point of suction in the tank?

2-19 A $\frac{3}{4}''$, 18 gauge copper condenser tube is used to condense steam on its outer surface at 5″ Hg pressure (absolute). Freon-12 is circulated through the tube at a saturated pressure of 25 psia and a quality of 40% at the tube entrance. The inside and outside coefficients (h_i and h_o, respectively) are 500 and 1500 BTU/hr-ft^2-°F. Determine the lbm/hr steam condensed per foot of tube length. If the Freon-12 exits at the end of a 10 foot section of the tube with a quality of 95%, what is the Freon-12 flow rate through the tube?

2-20 The Deacon process for oxidizing HCl is based on the reaction:

$$4\,HCl + O_2 \longrightarrow 2\,Cl_2 + 2\,H_2O$$

In a bench scale unit for studying this reaction, gaseous HCl and air, both at 25 °C, are mixed and fed to a catalytic reactor. The product gas leaves the reactor at 450 °C and has the following composition:

component	volume %
HCl	12.04
N_2	52.83
O_2	7.03
Cl_2	14.05
H_2O	14.05
	100.00

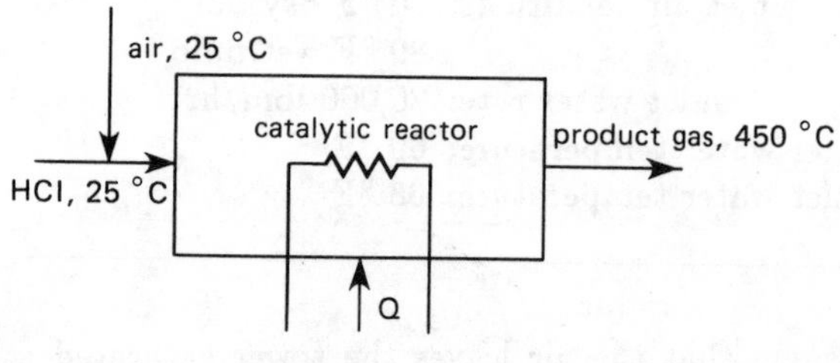

What is the net heat, Q, per 100 gmoles of product gas supplied to (or removed from) the reactor? Neglect heat losses to the surroundings.

Chemical Engineering Professional License Sample Exam # 3 A

3-1 Flue gas at 300 °F is fed to a smoke stack through a 4 foot diameter round duct. A pitot tube at the center of the duct and a manometer at its wall register 0.6 inches of water, gauge (iwg) and −0.5 iwg, respectively. The barometric pressure is 730 mm Hg, the gas contains 0.5 mole% sulfur dioxide, and the average molecular weight of the gas is 29. Assume all measurements are made well downstream of any flow disturbance. What is the emission rate of sulfur dioxide from the stack in lbm/hr?

3-2 A mixture of water, W, and acetic acid, A, is flash-distilled at atmospheric pressure. The feed consists of 70 mole% acetic acid and 30 mole% water.

$$\log_{10} p_A^0 = 7.188 - \frac{1416.7}{211 + t}$$

$$\log_{10} p_W^0 = 7.9668 - \frac{1668.2}{228 + t}$$

$$\log_{10} \gamma_A = 0.24 x_W^2$$

$$\log_{10} \gamma_W = 0.35 x_A^2$$

γ is activity coefficient, x is mole fraction, p^0 is vapor pressure in mm Hg, and t is temperature °C. (a) What is the bubble point? (b) Find the vapor and liquid flow rate and compositions if the feed is 100 lbmole/hr at 105 °C and 1 atm.

3-3 Continuous fractional distillation is used to separate a 1 lbmole/hr binary mixture, containing 47 mole% component A and 53 mole% component B, into the following three products:

- a distillate containing 90 mole% A
- an intermediate stream containing 75 mole% A
- a bottom product containing 10 mole% A

The flow rates of the distillate, intermediate, and bottom products are 0.3, 0.2, and 0.5 lbmole/hr, respectively. All of the products and the feed are at their boiling points, and the reflux ratio at the top of the tower is two. Additional reflux is provided on the intermediate product stage by taking off and condensing 0.4 lbmole/hr of vapor, splitting the condensate by removing 0.2 lbmole/hr as intermediate, and returning 0.2 lbmole/hr to the tray as reflux. The equilibrium data for the system are:

x, mole%	y, mole%
10	18.2
20	33.3
30	46.2
40	57.2
50	66.7
60	75.0
70	82.3
80	88.9
90	94.8

What are the equations of the operating lines? Sketch the equilibrium lines and the operating lines and show how you would determine the number of theoretical stages required.

3-4 Air at 10,000 SCFM containing 1000 ppm HCl is scrubbed in a packed tower with a recirculating water solution at about 40 °C and atmospheric pressure. The scrubber is designed to remove 95% of the HCl. The recirculating 60,000 lbm/hr water-HCl stream is maintained at a concentration of 5% HCl at the top inlet by bleeding off a slip stream of the HCl-water mixture to a waste neutralization plant and adding make-up water to the column as required. The column is packed with 2″ plastic Tellerettes. The expected H_{OG} for the system is 1.5 feet. Use the equilibrium data for HCl and water given to calculate the required packing height.

wt % HCl	p_{HCl}, mm Hg
2	5.5×10^{-4}
3	1.3×10^{-3}
4	2.3×10^{-3}
5	4.0×10^{-3}
6	6.5×10^{-3}
10	3.0×10^{-2}

3-5 The second-order reaction $2A \longrightarrow R$ is carried out in two continuously stirred tank reactors (CSTRs) in series. The reaction rate constant, K, is 10 and the reactor volumes are 3 and 12 gallons. The flow is 3 gpm, and initial concentration of A is 1 gmole/liter. Which combinations of reactors gives the largest conversion?

3-6 The triple-effect evaporator shown is used to evaporate water from raw potassium chloride solution. The hot concentrated brine is later purified in a vacuum crystallizer. For months the evaporator has operated at design capacity. The temperatures in the three effects were 140 °F, 105 °F, and 70 °F, respectively. During the last month the steam pressure in the first effect was increased from 115 psia to 135 psia in order to maintain design capacity. Steam consumption has increased 23% and the effect temperatures are now 140 °C, 119 °C, and 77 °C. The pressures in each effect correspond to the vapor pressure of the brine at the respective temperatures and compositions. The cooling water is within its usual temperature range at the condenser inlet and its flow rate is unchanged. The boiling point rise in each effect is about 5 °C. What is the reason for this costly development? Base your explanation upon analysis of the change in the heat transfer that has occurred, assuming that all temperature measurements are accurate.

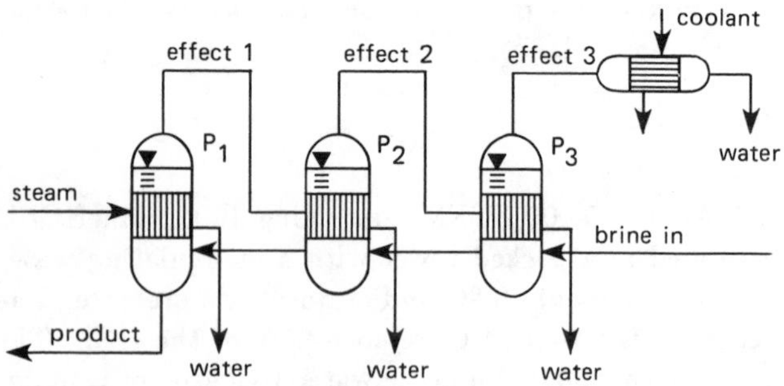

3-7 The system shown is constructed of 1″ schedule-40 screwed pipe and consists of one elbow, two valves, a heat exchanger, and lengths of pipe. The flow rate is 20 gpm. What is z?

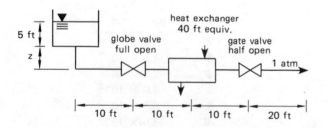

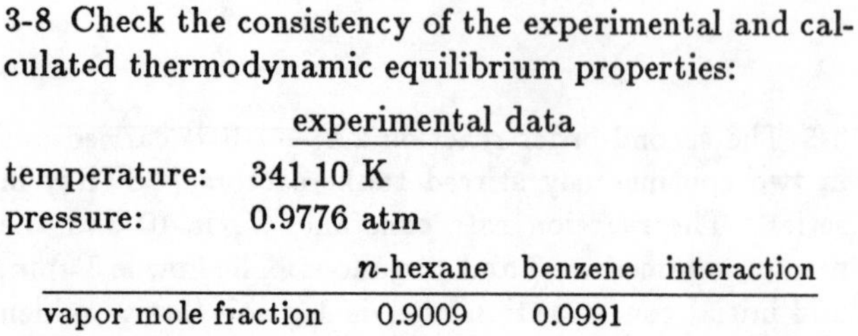

3-8 Check the consistency of the experimental and calculated thermodynamic equilibrium properties:

experimental data

temperature: 341.10 K
pressure: 0.9776 atm

	n-hexane	benzene	interaction
vapor mole fraction	0.9009	0.0991	
liquid mole fraction	0.9010	0.0990	

calculated from correlations

	n-hexane	benzene	interaction
vapor phase 2nd virial coefficient, cm^3/gmole	−1304	−1043	−1169
vapor pressure, atm	0.9808	0.6793	
activity coefficient in liquid phase	1.0033	1.3964	
liquid density, g/cm^3	0.617	0.830	
molecular weight	86.170	78.110	

3-9 A reactor vent gas and a polluted air stream are mixed and sent to a scrubbing tower for removal of the pollutants and water vapor. The reactor vent gas leaves the reactor at 203 °F and 500 cfm saturated with water. The polluted air stream is 1200 cfm at 80 °F and 50% relative humidity. The gas is scrubbed with a packed tower using the pumped recirculation system shown, which is used to minimize the amount of contaminated water to be discharged. A cooler is provided in the recirculating water stream to maintain the scrubber water temperature at 68 °F. Assume the gas leaving the scrubber is at 72 °F, saturated with water vapor, and the reactor is operated at atmospheric pressure. What is the water discharge rate?

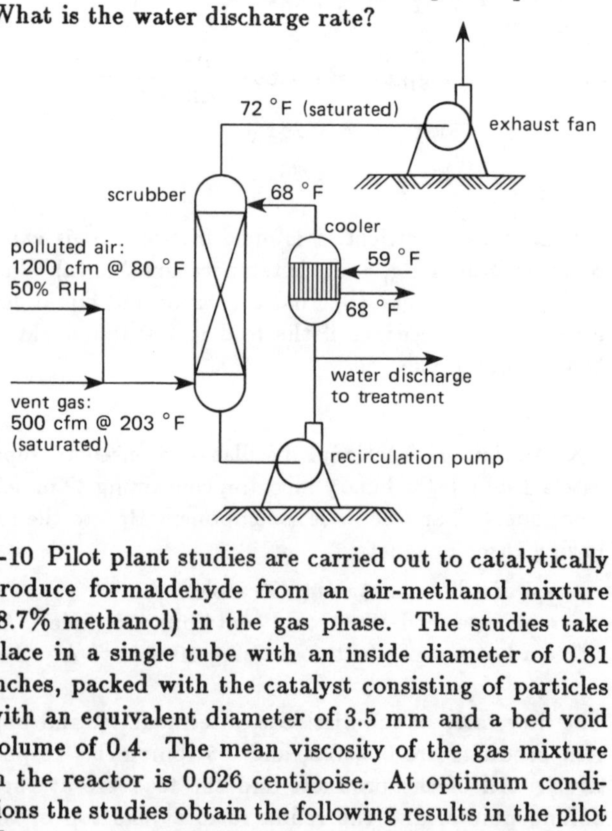

3-10 Pilot plant studies are carried out to catalytically produce formaldehyde from an air-methanol mixture (8.7% methanol) in the gas phase. The studies take place in a single tube with an inside diameter of 0.81 inches, packed with the catalyst consisting of particles with an equivalent diameter of 3.5 mm and a bed void volume of 0.4. The mean viscosity of the gas mixture in the reactor is 0.026 centipoise. At optimum conditions the studies obtain the following results in the pilot plant:

superficial gas mass velocity, G = 1330 lb/hr-ft^2
inlet temperature = 539 K
bed depth = 28.2 in
bed pressure drop = 0.12 atm
conversion = 91.4%

A multitubular reactor for commercial production will be designed using the same diameter tubes and having the same conversion and temperature profiles as a function of space-time as the pilot plant. Because of the relatively high pressure drop, a shallower bed must be used to reduce energy costs. A design criterion of 0.06 atm has been chosen. (a) What is the required superficial gas mass velocity, G, for the production unit? (b) What is the bed depth for the production unit?

Chemical Engineering Professional License Sample Exam #3 B

3-11 Analysis of the products of combustion of oil ($C_{12}H_{26}$) at a rate of 1000 lbm/hr with 20% excess air shows that there is no carbon monoxide present. These gases enter a superheater at 1550 °F and exit at 1150 °F. Steam enters the superheater at a pressure of 400 psia, a quality of 90%, and a flow rate of 10,000 lbm/hr. The mean specific heat of the products of combustion, c_p, is 0.25 BTU/lbm-°F. What is the condition of the steam (pressure, temperature, enthalpy) at the exit of the superheater?

3-12 Ammonia is formed by burning air and hydrogen to give a mixture of 3:1 H_2 to N_2 for the reactor. Air has the composition O_2:21%, N_2:78.006%, and Argon: 0.994%. Using the process shown, determine the composition of the recycle stream.

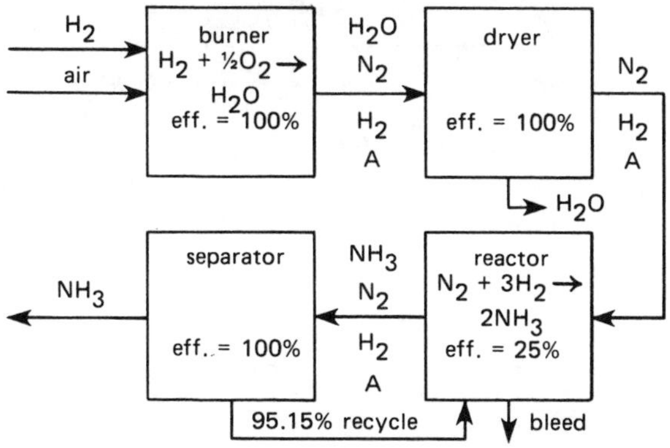

3-13 A wetted wall evaporating column produces the following data during a small-scale laboratory test. The rate of evaporation of *n*-butyl alcohol is predicted when operating the same equipment under the following conditions. Both experiments are carried out at 25.9 °C.

	water	*n*-butyl alcohol
total pressure, mm Hg	518	820
average gas flow rate, g/min	120	100
partial pressure of component evaporating, mm Hg	138	30.5
diffusivity, cm^2/s	0.258	0.087
vaporization rate, g/min	13.1	?

What is the rate of evaporation of *n*-butyl alcohol in the test?

3-14 A single-effect evaporator concentrates 10,000 lbm/hr of a 5% NaOH feed solution. The heat exchange area is 2500 ft^2, and 40 psig steam of 70% quality is available. If the feed enters at 70 °F and the evaporator operates at 1 psia, what steam rate is needed to make 65% NaOH?

3-15 A material containing 30% salt, 40% pulp, and 30% water is extracted in three stages with 0.5 lbm water/lbm of original feed used in each stage. The pulp shows no preferential adsorption of salt and retains 0.5 lbm solution/lbm of pulp. (a) What is the composition of the solid and liquid phase from each stage? (b) What is the composition of the dried (unwashed) leached solids after three stages? (c) How much lbm salt per 1000 lbm original feed is extracted?

3-16 A mixture of hydrocarbons is flash-distilled at 450 °C and 50 atm. The feed enters at 100 moles/hr; the vapor stream is 51.89 moles/hr. Feed compositions and vapor-liquid constant at 450 °C and 50 atm are given. Find the vapor and liquid compositions.

i	z_i	K_i
C_2H_6	0.002	16.2
C_3H_8	0.040	6.3
C_4H_{10}	0.332	1.35
C_5H_{12}	0.332	0.90
i-C_6H_{14}	0.152	0.92
n-C_6H_{14}	0.131	0.49
n-C_8H_{18}	0.011	0.09

3-17 The irreversible reaction shown is carried out in a back mixed reactor at 200 °F. Component A is fed at 0.5 lbmole/ft^3 and the reactor volume is 1 ft^3.

$$A \longrightarrow \text{products}$$

V, ft^3/min	C_A, lbmole/ft^3
0.2	0.030
0.4	0.057
0.8	0.102
1.6	0.170
3.2	0.253
6.4	0.336

What conversion will result if A is fed to a plug flow reactor with a concentration of 0.5 mole/ft^3 at 3 ft/min?

3-18 A forced draft cooling tower cools 1000 lbm/hr-ft^2 of water from 110 °F to 90 °F on a hot day when ambient air conditions are 120 °F and 70 °F wet bulb temperature. The outlet air at these conditions is 100 °F and 96 °F wet bulb. What is the final water temperature if the entering air on a cooler day is 100 °F and 50% relative humidity? Assume the enthalpy of saturated air can be represented by the equation:

$$h = a + bt + ct^2$$

where h = air enthalpy, BTU/lbmole dry air

$a = 2295$

$b = -63.85$

$c = 0.598$

$t \equiv$ temperature, °F

3-19 An elementary irreversible reaction gives 90% conversion in half an hour in a constant volume batch reactor with pure A. Assume A and P are ideal gases. How much time is required to get the same conversion in a constant pressure reactor?

$$2A \longrightarrow P$$

3-20 Find the second-stage raffinate composition and the percent component C extracted in the first stage for the simple multistage extraction process shown. An equilibrium diagram is provided.

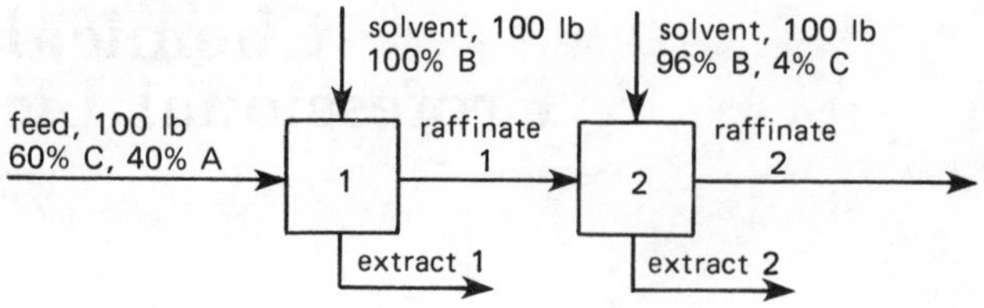

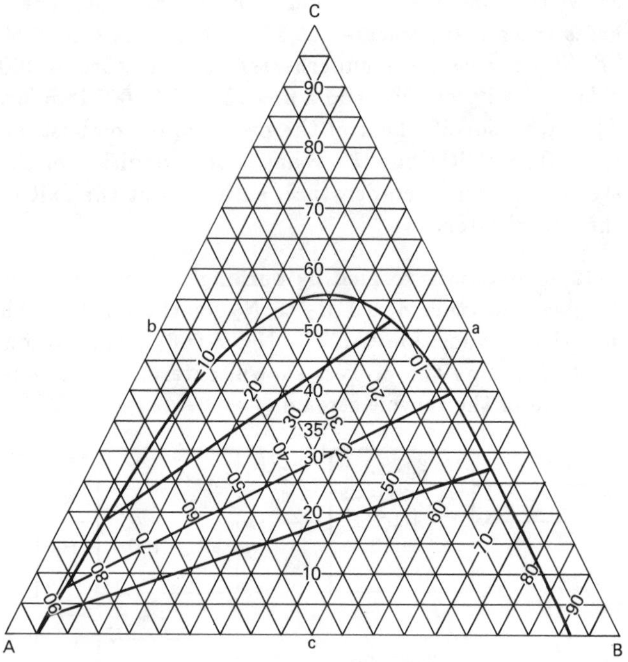

Chemical Engineering Professional License Sample Exam #4 A

4-1 An engineering firm must choose between two schemes for a pollution control facility with a project horizon of 10 years. In scheme I, a single plant (A) with a life of 10 years will be built. Scheme II involves building two identical plants (B and C), each with a life of five years. According to scheme II, plant C will replace plant B at the end of five years.

	scheme I	scheme II	
	A	B	C
capital cost ($)	25,500	13,000	19,500
installation cost ($)	10,000	6,000	5,000
salvage value ($)	2,000	1,500	2,500
hourly operational cost ($)	10.50	8.50	8.00
capacity, tons/hr	50	20	20

interest rate = 7%

(a) What is the yearly operational cost per ton for the two schemes over their expected lives? (b) What range of yearly tonnage rates will yield the lowest costs for the two schemes?

4-2 Sulfur dioxide is removed from a gas stream containing 13% SO_2 and 87% air (by volume) using a tower operated at atmospheric pressure. Water is used as the solvent. The gas enters the tower at 100 cfm at 40 °C and 1 atm pressure. The tower is maintained at 40 °C throughout, and the desired removal is 99%. What is the minimum water required?

4-3 A mixture of dry air and hydrogen chloride at 20 psia and 20 °C is passed through a tank containing 250 ml of 0.01 N NaOH solution. The residual NaOH is back-titrated with 17.49 ml of 0.1 N HCl. The dry air has a measured volume of 1000 cc at a pressure of 750 mm Hg at 25 °C, measured in a wet test meter. Calculate the mole fraction and mass fraction of HCl in the original mixture.

4-4 A double-effect, feed-forward evaporator concentrates a 15% aqueous solution to a 60% solution. Feed enters at 80 °F, and the 60% solution leaves at 140 °F. Steam for the first effect is at 20 psia. Overall heat transfer coefficients for the first and second effect are 500 and 250 BTU/hr-ft^2-°F, respectively. Set up the iterative calculation and show the revised values to be used in the next iteration.

4-5 An industrial process requires the addition of sulfuric acid to water in a tank. The tank has a 600 gallon capacity, and 50 gallons of 98% H_2SO_4 are added to 500 gallons of water. The resultant solution is to be cooled to 80 °F before use. Both liquids are initially at 70 °F. The heat exchanger is a double-pipe countercurrent type using water in the annulus entering at 50 °F and leaving at 70 °F. The overall heat transfer coefficient, U_o, is 500 BTU/hr-ft^2-°F. The tank can be emptied in 30 minutes. What is the required heat exchanger surface area?

4-6 Use the reaction and data provided to investigate equilibrium. What conclusions can you draw about the possibility of its industrial application?

$$N_{2(g)} + C_2H_{2(g)} \rightleftharpoons 2\,HCN_{(g)}$$

(Units used are gcal, gmoles, and K.)

	N_2	C_2H_2	HCN
H_f at 25 °C	0.00	+53,900	+30,800
entropy at 25 °C	45.79	47.5	48.25

4-7 A pump suction line is composed of 22 feet of 10″ diameter cast iron pipe and has one 90° elbow and one wide open gate valve. Suction lift is 15 feet. The pump operates at 3000 feet altitude, where the pressure is 29.9″ Hg. What is the net positive suction head, NPSH, to the pump if the throughput is 1000 gpm?

4-8 The dimerization of butadiene according to the equation:

$$2\ C_4H_6 \longrightarrow C_8H_{12}$$

is described by the rate equation:

$$-r_A = kC_A^2$$

k = 0.85 liter/gmole-min at 326 °C. This reaction will be carried out in a plug flow reactor, PFR, operated at 10 atm constant pressure (absolute) and at a temperature of 326 °C. The feed is composed of 20% butane and 80% butadiene at a design rate of 20,000 lbm/day of butadiene feed with 90% conversion. Butane is inert. All gases act as ideal gases at these conditions. Determine the reactor volume for this condition.

4-9 The rate constant, k, for a reaction has been determined as a function of temperature as follows:

t °C	k, sec^{-1} ($\times 10^5$)
0	1.20
15	8.34
30	46.50
50	360.00

Calculate the activation energy and the pre-exponential factor for the reaction.

4-10 A finned tube heat exchanger in a warehouse heats air with process steam at a total pressure of 1 atm. A $1\frac{1}{2}''$ schedule-40 steel pipe is used with enough fins to give 3 ft^2 of area per foot of length (2.5 ft^2 for fins, 0.5 ft^2 for the pipe). The steam quality is 0.95, and the condensate exits at 200 °F. (a) How much steam is required if 60,000 BTU/hr are needed to maintain the warehouse at 70 °F? (b) What length of pipe is necessary if the air temperature rise in the exchanger is 30 °F?

$$h_{\text{air}} = 5 \text{ BTU/hr-ft}^2\text{-°F}$$
$$h_{\text{steam}} = 200 \text{ BTU/hr-ft}^2\text{-°F}$$
$$\text{fin efficiency} = 80\%$$
$$(c_p)_{\text{air}} = 0.25 \text{ BTU/lbm-°F}$$

Chemical Engineering Professional License Sample Exam #4 B

4-11 A construction firm has compiled data on the cost of operating a bulldozer. The data given are in thousands of dollars, and the interest rate is 10%.

year	item	amount
0	initial cost	\$24,000
1	operation and maintenance	6,000
2	operation and maintenance	7,500
3	operation and maintenance	9,000
	overhaul	7,400
	salvage value	8,000
4	operation and maintenance	7,200
5	operation and maintenance	8,200
6	operation and maintenance	9,200
	overhaul	10,400
	salvage value	6,000
7	operation and maintenance	8,800
8	operation and maintenance	9,900
	salvage value	0

Compare the Equivalent Uniform Annual Cost (EUAC) of operating the bulldozer for three, six, and eight years.

4-12 An acid stream consisting of 40% H_2SO_4, 30% HNO_3, and 30% H_2O at 20,000 lbm/hr must be produced. Three streams are added to a mixer to produce the stream: 98% H_2SO_4, 80% HNO_3, and a waste acid stream consisting of 5% HNO_3 and 20% H_2SO_4. What are the flow rates of the three streams?

4-13 A first-order irreversible reaction, A $\longrightarrow$ B, has a reaction rate constant, k, of 0.42 min^{-1}. The reaction occurs in a continuously stirred tank reactor (CSTR) ($\tau = 2$ min) initially filled with an inert liquid. At $t = 0$, pure A is fed, and an overflow is withdrawn. At what time will the concentration of component B be at 99% of the steady state value?

4-14 A mixture of water and acetic acid is flash-distilled at atmospheric pressure. The feed consists of 70 mole% acetic acid and 30 mole% water. The following data are given:

$$\text{acetic acid: } \log_{10} p_a^0 = 7.188 - \frac{1416.7}{211 + t}$$

$$\log_{10} \gamma_a = 0.24\, x_W^2$$

$$\text{water: } \log_{10} p_W^0 = 7.9668 - \frac{1668.2}{228 + t}$$

$$\log_{10} \gamma_W = 0.35\, x_A^2$$

γ is the activity coefficient, x is the mole fraction, p^0 is the vapor pressure in mm Hg, and t is the temperature in °C. (a) What is the bubble point? (b) Find the vapor and liquid flow rate and compositions if the feed is 100 lbm mole/hr at 105 °C and 1 atm.

4-15 A company produces 60% NaOH from a 10% solution using a single-effect evaporator. A feed rate of 2000 lbm/hr is supplied. The vapor space pressure is 1 psia and steam at 100 psig is available. The overall heat transfer coefficient, U_o, is 60 – 80 BTU/hr-ft^2-°F. Two evaporators are available, one with 80 ft^2 and the other with 400 ft^2. Which evaporator would you use?

4-16 A plant produces soap as a 50% aqueous solution at a rate of 1 million lbm/yr (100% basis). The soap is sold for \$0.30/lbm of solution FOB plant. The freight paid by most customers is \$0.028/lbm of solution. Customers usually dilute the solution to 10% before use. It has been determined that the 50% solution could be concentrated to 90% in a single-stage evaporator with a steam consumption of 1.2 lbm steam/lbm water evaporated. Steam cost is \$0.003/lbm and no additional labor is required. Annual capital related costs (depreciation, maintenance, taxes, insurance, etc.) are 20% of total capital. Since the 90% solution would reduce freight costs, customers would pay more for the soap content of the 90% solution without increasing their overall costs. The design constraint for the new concentrating equipment requires that the net cost to the customer must not be increased. What is the maximum total investment that could be made, assuming that the minimal acceptance rate of return is 30% per year before taxes?

4-17 A granular solid with a bulk density of 90 lbm/ft^3 is dried in a batch dryer with air at 150 °F and a humidity of 0.005 lbm H_2O/lbm dry air. The solid contains 0.5 lbm H_2O/lbm dry solid and is placed in 1″ deep insulated pans so that transfer occurs only on the top surface. The final moisture content is 0.02 lbm H_2O/lbm dry solids, and the critical moisture content is 0.1 lbm H_2O/lbm dry solids. Air is passed over the pans at 1200 lbm/hr-ft^2. (a) What drying time is required? (b) What is the drying time if the air temperature is raised to 200 °F?

4-18 Hot water enters a single seat control valve as a saturated liquid at 2000 psia. The valve operates at a back pressure of 1000 psi and has a flow coefficient, C_v, of 400. (a) What is the mass flow rate through the valve with no flashing? (b) What is the mass flow rate through the valve corrected for flashing?

4-19 A non-ideal binary mixture of benzene and cyclohexane forms an azeotrope at 760 mm Hg and mole fraction of benzene, $x_b = 0.55$ at 77 °C. The vapor pressures of the pure compounds are:

$$\text{benzene:} \quad \ln p_b = 17.61922 - \frac{3880.00}{T}$$

$$\text{cyclohexane:} \quad \ln p_c = 17.40484 - \frac{3811.61}{T}$$

p is in mm Hg, and T is in K. (a) What are the coefficients of the van Laar equation? (b) What is the vapor composition of $x_b = 0.2$ at a total pressure of 1 atm?

4-20 A saturated liquid feed containing 42 mole% heptane and 58 mole% ethyl benzene is fractionated at 760 mm Hg to produce 97 mole% heptane and a residue containing 99 mole% ethyl benzene. What is the number of equilibrium stages required at a reflux ratio of 2.5? The Antoine constants (p in mm Hg) are:

	heptane	ethyl benzene
A	6.9024	6.9537
B	1268.12	1421.91
C	216.9	212.9

Chemical Engineering Professional License Sample Exam #5 A

5-1 A 100% recirculating air dryer system receives air from a cooler at 34 °F. The air is heated to 180 °F and is passed through a process dryer. The air leaves the first dryer in a saturated condition and is reheated to 174 °F, passes through a second dryer, exits at 80% relative humidity, and is passed to the cooler. (a) How many pounds of water are evaporated in each dryer? (b) What is the heating duty of each heater? (c) How many cubic feet of air leave each dryer? Express values in pounds of dry air.

5-2 A packed tower, which performs adequately, scrubs a small amount of a noxious substance from an air stream with water. An expansion will produce an increase of 60% in the volume of the air stream, but there will be no change in the concentration of the noxious element. The tower is packed with 2″ Raschig rings, and is 64″ in diameter with a packed height of 14 feet. The water flow is 440 gpm. Air enters at 70 °F and 0.5 psig and discharges at atmospheric pressure. If the water flow is increased by 60%, will the tower still perform its function of removing the noxious element?

5-3 A plant process stream of 150 gpm of aqueous solution carries minute solids that are deposited out in an evaporator downstream. These solids will be removed using a leaf filter with 20 leaves, 3 feet in diameter. A piece of filter cloth is strapped over a 6″ diameter funnel and experiments are performed with the slurry. In 2.1 minutes, 0.32 gallons of filtrate are collected, and in 12 minutes, 0.97 gallons of filtrate are collected. The tests are conducted under vacuum conditions (11 psia), and the cake is incompressible. The leaf filter will operate at 30 psia until the filtration rate drops to 10% of the clean filter rate; then the filter will be cleaned, an operation that takes eight minutes. Will the filter do the job?

5-4 A liquid mixture containing equimolar amounts of three compounds is pumped through a long pipe at 5 atm pressure and a constant temperature of 100 °F. Using the vapor-liquid equilibrium data given, find the maximum pressure drop in the pipe before the liquid vaporizes.

$K_i = \alpha_i (T/p)$ where T is in °F; p is in atm

$\alpha_1 = 0.02$

$\alpha_2 = 0.03$

$\alpha_3 = 0.05$

5-5 Acetic anhydride (AA) is reacted with water to form acetic acid. The concentration of AA is low (1.5% by weight). The reaction is first-order, and the reaction rate constant is given by the relation $\ln k = [A - B/T]$, where T is in K, A = 8.00, and B = 2928. The reactors used are two continuously stirred tank reactors (CSTRs) in series, both having the same volume. The throughput is 50 gpm. (a) Given a 75% conversion at 25 °C, calculate the reactor volumes. (b) Given a conversion of 90%, calculate the required temperature of reaction using the same reactor volumes as part (a). (c) Given a conversion of 90% at 25 °C, calculate the required reactor volumes.

5-6 Two oil streams are treated with an absorber/stripper. Heat exchangers are being considered to provide additional heat. The specifications are:

	flow rate lbm/hr	temp in °F	temp out °F
rich stream	11,000	80	212
lean stream	10,000	290	145

Data for both streams are:

$$\text{density} = 50 \text{ lbm/ft}^3$$

$$c_p = 0.4 \text{ BTU/lbm-°F}$$

$$\mu = 5 \text{ centipoise at 80 °F}$$

An old heat exchanger ($A_o = 100 \text{ ft}^2$, two tube passes, clean) is given two field trials:

(1) rich oil in the shell, steam at atmospheric pressure, and at steady state

$$\text{flow} = 11,000 \text{ lbm/hr}$$

$$\text{temperatures} = 80 \text{ °F (in); } 210 \text{ °F (out)}$$

(2) lean oil in the tube, steam at atmospheric pressure, and at steady state

$$\text{flow} = 10{,}000 \text{ lbm/hr}$$

$$\text{temperatures} = 80\ ^\circ\text{F (in); } 211\ ^\circ\text{F (out)}$$

Assume no fouling of the heat exchanger. Is the heat exchanger adequate for either stream?

5-7 The first-order reaction A $\longrightarrow$ R is carried out in a continuously stirred tank reactor (CSTR). The conversion is 75% and the reaction occurs at constant density. It is proposed to convert the CSTR to a batch reactor having an eight minute downtime between batches. (a) What is the conversion of A for the batch reactor, assuming that the volumes of the two reactors are equal and all other conditions remain the same? (b) Determine the percent increase in production using the batch reactor. What are the drawbacks, if any, to this change?

5-8 Dry air enters the inner tube of a vertical double-pipe heat exchanger at 300 °F, 30 psia, and a velocity of 100 ft/sec in the tube. The air flows at 200 lbm/hr and leaves the heat exchanger at 0 °F, 15 psia, and 10 feet above the entrance. What is the total heat removed in BTU/hr if the heat capacity of the air follows the expression $c_p = a + bT + cT^2$, where $a = 6.39$, $b = 9.8 \times 10^{-4}$, $c = -8.18 \times 10^{-8}$, T is in °R, and c_p is in BTU/lbmole-°R?

5-9 Styrene monomer is polymerized (40% conversion) in a continuously stirred tank reactor (CSTR). The resulting mixture is cooled in a heat exchanger to 180 °F with cooling water entering at 45 °F countercurrently. The monomer feed rate is 941 g/s. Find (a) the heat of reaction per mole of monomer consumed and (b) the volume of the reactor, given the following data:

reaction rate:

$$-r_{\text{monomer}} = 0.869 \text{ moles/liter-sec}$$
$$A_o = 500 \text{ ft}^2$$
$$U_o = 100 \text{ BTU/hr-ft}^2\text{-}^\circ\text{F}$$
$$c_{p\text{mixture}} = 0.45 \text{ cal/gmole-}^\circ\text{F}$$
$$MW_{\text{monomer}} = 104 \text{ g/gmole}$$
$$MW_{\text{polymer}} = 140{,}000 \text{ g/gmole}$$
$$C_{Ao} = 0.9423 \text{ moles/liter}$$

exit cooling water = 180 °F

5-10 In a reaction chamber n-butane is burned to CO_2 and CO using 100% excess air. The gases enter the chamber at 25 °C and leave at 851 °C. What is the percent n-butane burned to CO_2?

$$c_p = a + bT + cT^2 \quad \text{cal/mole-}^\circ\text{C} \quad (T \text{ in K})$$

	a	$b \times 10^3$	$c \times 10^6$	H_f kcal/mole
n-C_4H_{10}	8.4228	0.9739	−0.3555	−29.812
O_2	6.0954	3.2533	−1.0171	0
CO_2	6.3930	10.1000	−3.4050	−94.0518
CO	6.324	1.863	−0.2801	−26.4157
$H_2O_{(g)}$	7.219	2.374	0.267	−57.798
$H_2O_{(l)}$	18.03	0	0	−68.3174
N_2	6.442	1.4125	−0.0807	0

Chemical Engineering Professional License Sample Exam #5 B

5-11 Water is supplied at 50 gpm to a pressure vessel operating at 20 psig and 80 °F. A surge tank, vented to the atmosphere, is provided in the figure shown. During normal operations the check valve is open. The equivalent length of the stand pipe to the pressure vessel is 100 ft and includes all elbows, fittings, and expansion. (a) What is the level of water in the surge tank under normal flow conditions? (b) If the pump shuts off suddenly and the check valve closes immediately, how long will it take for the flow to the pressure vessel to decrease by 5%?

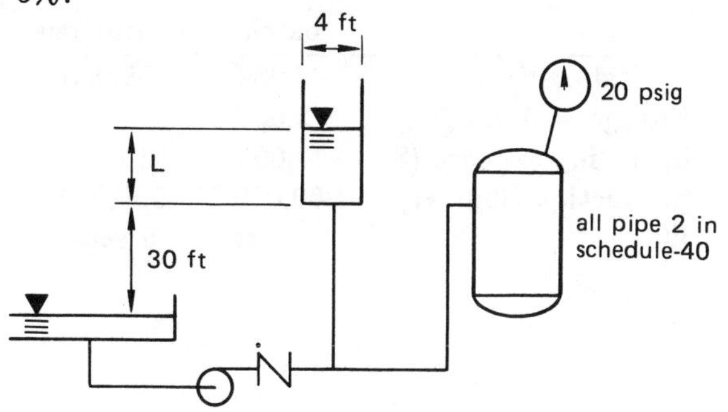

5-12 Water gas (30% CO and 70% H_2) reacts catalytically at elevated temperatures according to the reaction:

$$CO_{(g)} + H_2O_{(g)} \rightleftharpoons CO_{2(g)} + H_{2(g)}$$

Assume 2 moles of steam per mole of CO at 425 °C and atmospheric pressure. The equilibrium constant for the reaction is:

$$\ln K = \frac{5038.8}{T} - 0.09415 \ln T + 0.0014553\, T - 2.5 \times 10^{-6} T^2 - 5.269$$

T is measured in degrees Kelvin. How much hydrogen is produced per mole of CO-H_2 mixture of feed at equilibrium?

5-13 A continuous countercurrent extraction battery is used to extract uranyl nitrate, $UO_2(NO_3)_2$, in water with tributyl phosphate, an organic. Water and tributyl phosphate are essentially immiscible. The aqueous stream originally containing the uranyl nitrate flows at 100 liters/min, and the organic stream flows at 150 liters/min. The molar distribution of uranyl nitrate is constant at 10 organic/aqueous. How many stages of extraction are required to reduce the aqueous stream from 0.5 moles uranium/liter to 1 ppm?

5-14 A single-effect evaporator is used to concentrate a feed of 40,000 lbm/hr of 20% NaOH entering at 110 °F. Saturated steam is available at 120 psig. The pressure in the evaporator will be maintained at 2″ Hg absolute. The area of the evaporator is 460 ft^2, and the heat transfer coefficient is 300 BTU/hr-ft^2-°F. The unit will operate at a liquid temperature of 175 °F. Calculate the water evaporated and the concentration of the evaporator product. The specific heat of solutions of NaOH is given by the relation:

$$c_p = 1.00 - 0.032\sqrt{\text{weight}\%}$$

Assume that the boiling point elevation (BPE) of a 20% NaOH solution is 10 °F, the BPE at 40% is 40 °F, and the BPE varies linearly.

5-15 The data given are for a constant pressure filtration of a $CaCO_3$ slurry in H_2O. The filter is a 6″ filter press with an area of 1.0 ft^2. The mass fraction of solids in the feed is 0.139; the mass ratio of wet cake to dry cake is 1.59 in the experiment when the pressure is 5 psig and 1.47 at the other pressures; the dry cake density is 63.5 lbm/ft^3 in the run at 5 psig, 73.0 lbm/ft^3 in the runs at 15 and 30 psig, and 73.5 lbm/ft^3 in the run at 50 psig. The data are used to calculate the filtration time and the volume of filtrate for a larger filter press operated at 25 psig, having a 10 ft^2 filter area and a frame thickness of $1\frac{1}{2}''$. What are the one-cycle filtration time and filtrate volume of the larger filter? How long will it take to wash the filter cake with a volume of wash water equal to the one-cycle filtrate volume at 80% of the theoretical rate?

filtrate lbm	5 psig time sec	filtrate lbm	15 psig time sec	30 psig time sec	50 psig time sec
0	0	0	0	0	0
2	24	5	50	26	19
4	71	10	181	98	68
6	146	15	385	211	142
8	244	20	660	361	241
10	372	25	1009	555	368
12	524	30	1443	788	524
14	690	35	2117	1083	702
16	888				
18	1188				

5-16 Water vapor leaving a single-effect evaporator is condensed using water at 70 °F. The evaporator produces 24,000 lbm/hr of vapor from a solution having a negligible BPE. The condenser has an area of 665.4 ft^2. Experience has shown that 48,000 lbm/hr of cooling water can be used and that the overall heat transfer coefficient is 500 BTU/hr-ft^2-°F. Find the condensing temperature and the pressure in the evaporator.

5-17 What is the volume of the plug flow reactor used for the non-elementary homogeneous gas phase reaction A $\longrightarrow$ 3R having the rate $-r_A = kC_A^2$? The feed consists of 50% inerts and 50% A at 200 °C and 5 atm, and is flowing at 1 liter/sec. The initial concentration of A is 0.0625 moles/liter, and the required conversion is 90%. The reaction rate constant is 0.01 liter/mole-sec at 200 °C.

5-18 A porous solid is dried in a batch dryer under constant drying conditions. Six hours are required to reduce the moisture content from 30% to 10%. The critical moisture content is 16%, and the equilibrium moisture content is 2%. How long will it take to dry a sample of the same solid from 35% to 6% under the same drying conditions? All moisture contents are on a dry basis. Assume that the rate of drying during the falling rate period is proportional to the free-moisture content.

5-19 An air stream at atmospheric pressure is continuously cooled from 38 °C to 15 °C with an air conditioner. The amount of air cooled is 30,000 liter/min (measured at 1 atm and 25 °C). The temperature of the air to which the air conditioner is discarding heat is 38 °C. What is the minimum power requirement of the air conditioner?

5-20 A company is considering two alternatives for the production of polyols, one batch and one continuous. Polyols sell for \$0.10/lbm. Assume a tax of 48%. Which alternative should be selected if the after-tax rate of return must be at least 15%? Straight-line depreciation is used.

	batch	continuous
capital cost (\$)	27,000	55,000
salvage at 5 yrs (\$)	2,000	1,900
operating expense (\$)	20,000	19,000
production (lbm/yr)	960,000	995,000
life	5 years	5 years

Chemical Engineering Professional Sample License Exam #6 A

6-1 A centrifugal pump is used to pump water at 45 °F from a river to a closed tank held at atmospheric pressure by the control valve shown in the diagram. The pipe is 2″ schedule-40 and has a total equivalent length of 220 feet, including all fittings and entrance and exit effects. (a) What is the flow rate in gpm at steady state? (b) What size electric motor at 80% efficiency is needed? A pump chart is provided.

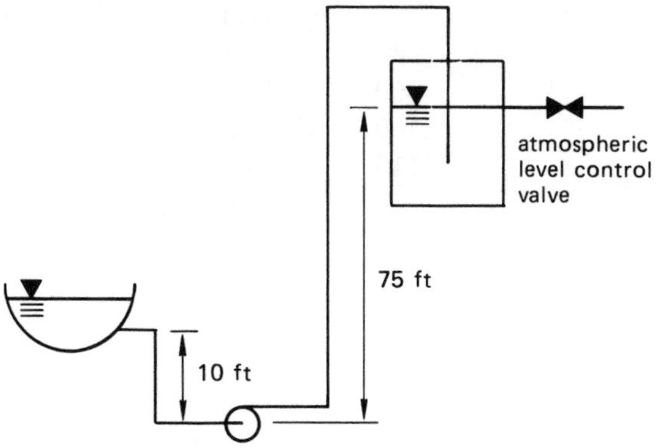

gpm	head, ft	efficiency
0	210	100%
10	205	94
20	200	88
30	195	82
40	190	76
50	183	70
60	175	65
70	162	60
80	146	56
90	130	52
100	112	49
110	90	46
120	65	44
130	25	42

6-2 A perfectly mixed tank containing 400 lbm of a liquid at 200 °F and specific heat of 0.94 BTU/lbm-°F is allowed to flow out through a control valve at the bottom of the tank at a constant flow rate of 80 lbm/hr the instant the same liquid enters the tank at the top at 100 lbm/hr and 80 °F. Assuming that the tank has enough capacity and will not overflow, what is the temperature of the liquid leaving the tank after six hours?

6-3 The autocatalytic reaction given takes place in a batch reactor. The reaction is elementary and has a rate constant of $k = 1.0$ liter/mole-min.

$$A + P \longrightarrow 2P$$

(a) If the batch initial concentration consists of 0.99 moles of component A and 0.01 moles of component P, and the desired product is to consist of 10% A and 90% P, what is the maximum rate? (b) At what time is the reaction rate at maximum? (c) How long will the reaction take?

6-4 A single-effect evaporator is used to concentrate 20,000 lbm/hr of 20% NaOH to 50% NaOH under a condenser pressure of 2 psia. Steam is used at 350 °F in the steam chest. The BPE of 50% NaOH is 50 °F, the temperature of the feed is 200 °F, and the overall heat transfer coefficient is 300 BTU/hr-ft^2-°F. (a) What is the required heat exchange area for the evaporator? (b) What is the steam economy?

6-5 A mixture of 90% methane, 6% ethane, and 4% nitrogen is burned with 30% excess air. What is the adiabatic flame temperature if the air is at 90 °F?

6-6 What is the theoretical steam rate of an engine (ideal and frictionless) when it is supplied with steam at 385 psig superheated to 600 °F and exhausted at 15.3 psig? What is the actual steam rate for an engine rated at 75% efficiency?

6-7 Phosphine at 1200 °F with a first-order rate decomposes according to the following reaction:

$$4\,PH_{3\,(g)} \longrightarrow P_{4\,(g)} + 6\,H_2$$

What size plug flow reactor operating at 1200 °F and 4.6 atm will produce 80% conversion of feed consisting of 4 lbmole/hr of pure phosphine with $k = 10$/hr?

6-8 Methyl alcohol flowing in the inner pipe of a double-pipe heat exchanger is cooled from 150 °F to 100 °F with 15,000 lbm/hr of 80 °F water. The inner pipe is made of 1″ schedule-40 pipe. The pipe has a thermal conductivity of 26 BTU/hr-ft²-°F. The allowed approach temperature is 15 °F. Using the individual coefficients and fouling factors given, determine the exchanger area required.

$$c_p = 0.65 \text{ BTU/lbm-°F (alcohol)}$$
$$h_i = 180 \text{ BTU/hr-ft}^2\text{-°F}$$
$$h_o = 300 \text{ BTU/hr-ft}^2\text{-°F}$$
$$h_{di} = 1000 \text{ BTU/hr-ft}^2\text{-°F}$$
$$h_{do} = 500 \text{ BTU/hr-ft}^2\text{-°F}$$

6-9 A mixture of 2 mole% ethanol and 98 mole% water is stripped in a plate column to a bottom product containing not more than 0.01 mole% ethanol. Steam is admitted through an open coil in the liquid on the bottom plate to provide the vapor for the column. The feed is at its boiling point and the steam flow is 0.2 moles/mole of feed. The equilibrium for dilute ethanol-water is a straight line, $y = 9.0x$. How many ideal plates are required?

6-10 The data given are obtained from a constant rate filtration of a sludge consisting of $MgCO_3$ and H_2O. The rate is 0.1 lbm/ft²-sec, the viscosity of the filtrate is 0.92 centipoise, and the concentration of solids in the feed is 1.08 lbm/ft³ of filtrate. (a) What is the resistance of the filter medium? (b) What are the cake compressibility exponent and the pre-exponent coefficient?

time, sec	Δp, psi
10	4.4
20	5.0
30	6.4
40	7.5
50	8.7
60	10.2
70	11.8
80	13.5
90	15.2
100	17.6
110	20.0

Chemical Engineering Professional License Sample Exam #6 B

6-11 A steam Rankine cycle operates with 100 psia saturated steam that is reduced to 1 atm through expansion in an 80% efficient turbine. The working fluid is 80 °F liquid at 1 atm upon entering a 60% efficient pump. What is the cycle thermal efficiency?

6-12 A vapor mixture at 1 atm and 100 °C contains 50 mole% propane, 30 mole% *n*-butane, and 20 mole% *n*-hexane. (a) What is the pressure required at a dew point of 100 °C? (b) What is the composition of the liquid at the dew point? (c) What is the pressure required for complete condensation at 100 °C?

6-13 Consider the irreversible second-order reaction and following data:

$$A + B = R + S$$

$$\text{flow} = 8\ \text{ft}^3/\text{min}$$
$$\text{reactor volume} = 2\ \text{ft}^3\ \text{(CSTR)}$$
$$C_{Ao} = 2\ \text{lbmoles/ft}^3$$
$$C_{Bo} = 3\ \text{lbmoles/ft}^3$$
$$X_a = 60\%$$

Assume constant density and no temperature change in the reactor. (a) What will be the overall conversions of A (X_A) if a second identical reactor is added in the series? (b) What will be the composition of the effluent from the second reactor?

6-14 A countercurrent heat exchanger is designed to heat 125 gpm water from 60 °F to 120 °F (maximum 130 °F) with a heat transfer coefficient, $U_{\text{dirty}} = 220$ BTU/hr-ft^2-°F, including a dirt factor of 0.001 using steam at 30 psia. (a) What is the exit water temperature when the heat exchanger is first operated (clean)? (b) What is the controlling parameter in order to keep the exit temperature less than 130 °F?

6-15 The piping system shown flows at 2500 gpm. What are the flows at points A and B? What is the pressure at point 1 when the pressure at point 2 is 50 psig?

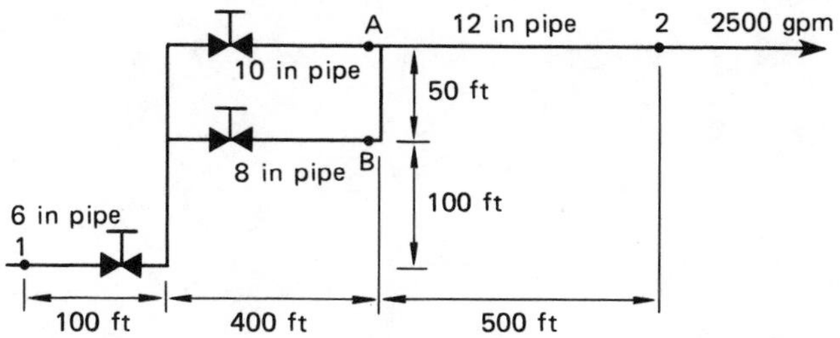

6-16 A non-elementary gas phase reaction is evaluated as a function of time:

$$A = R + S$$

p, mm Hg	*t*, minutes
340	0.00
357	0.36
391	1.26
408	1.84
425	2.52
459	4.30
475	5.48
510	8.71
595	30.75
663	233.36

The reaction is carried out at 270.1 °C. Find the reaction rate constant and order.

6-17 What is the heat loss per lineal foot from a 4″ schedule-40 cast iron pipe insulated with 1″ of 85% magnesia insulation? The pipe transports a fluid at 500 °F with an inner heat transfer coefficient, h_i, of 50 BTU/hr-ft^2-°F. The insulation surface is exposed to ambient air at 70 °F with an outer heat transfer coefficient, h_o, of 5 BTU/hr-ft^2-°F.

6-18 Use the pyridine-chlorobenzene-water equilibrium data given at 25 °C for the tie-line end points to determine the amount of solvent (chlorobenzene) required to reduce a 2000 pound batch of 50% aqueous pyridine to 2% in a single stage batch extraction process.

C_5H_5N	C_6H_5Cl	H_2O	C_5H_5N	C_6H_5Cl	H_2O
0.00	99.50	0.50	0.00	0.08	99.92
11.05	88.28	0.67	5.02	0.16	94.82
18.95	79.90	1.15	11.05	0.24	88.71
24.10	74.28	1.62	18.90	0.38	80.72
28.60	69.15	2.25	25.50	0.58	73.92
31.55	65.58	2.87	36.10	1.85	62.05
35.05	61.00	3.95	44.95	4.18	50.87
40.60	53.00	6.40	53.20	8.90	37.90
49.00	37.80	13.20	49.00	37.80	13.20

6-19 Given initial stoichiometric ratios for a reaction conducted at 500 °C and the data given, (a) what is the heat of reaction at 500 °C and (b) is heat added or removed at 500 °C? Heats of formation are given in cal/gmole and heat capacities are given in cal/g-K.

$$2\,CH_4 + 2\,Cl_2 \longrightarrow 2\,CH_3Cl + 2\,HCl$$

component	heat of formation at 25 °C	heat capacity T in K
CH_4	−17,899	$5.34 + 0.0115\,T$
Cl_2	0	$8.28 + 0.00056\,T$
CH_3Cl	−19,600	$6.26 + 0.0094\,T$
HCl	−22,063	$6.70 + 0.00084\,T$

6-20 A company is evaluating the alternatives of renting a car for four years or buying the car and selling it after the same length of time. The rental will cost $0.28 per mile with a usage of 12,000 miles per year. The car will cost $9000 now, with a yearly cost of $800 and a salvage value of $2000 after four years. If interest is 15%, taxes are 40%, and the depreciation allowances for years 1, 2, and 3 are 25%, 38%, and 37%, respectively, which alternative is better? Why?

Chemical Engineering Professional License Sample Exam #7 A

7-1 Find the enthalpy change of nitrogen gas (N_2) as it goes from 1 atm and 25 °C to 25 atm and 700 °C.

7-2 A bed of $\frac{1}{4}''$ cubes is used as packing in a regenerative heater. The cubes are poured into the cylindrical regenerator shell to a depth of 10 feet. Air enters the bed at 80 °F and 100 psia, and leaves at 400 °F, flowing at 1000 lbm/hr-ft^2. Use the Ergun equation to find the pressure drop across the bed.

7-3 The pump in the figure operates at 150 gpm and 170 °F. The suction line is 6″ schedule-40, and the entire discharge line is 4″ schedule-40 screwed pipe and fittings. The fluid is water. The suction friction loss is 1.8 ft/100 ft equivalent pipe, and the pressure friction loss is 8.5 ft/100 ft equivalent pipe. What is the total dynamic head (TDH) of the pump?

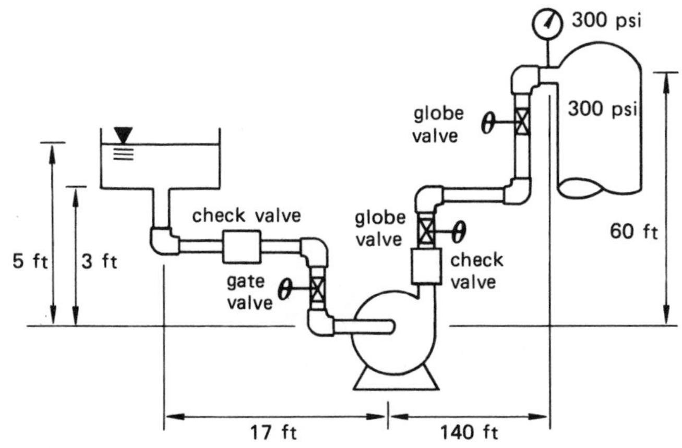

7-4 A sieve tray column is used to separate a feed stream of 100 lbmoles/hr of a mixture of 60 mole% benzene and 40 mole% toluene. The column has 16 actual plates, a partial condenser, and a partial reboiler. The feed stream is saturated vapor. The overall column efficiency is 50%. Due to loading problems, the reflux rate is 1.5 times the minimum reflux. The distillate required is 95% benzene. What is the maximum distillate rate achievable if the relative volatility is 2.2?

7-5 Octane (C_8H_{18}) is burned with 100% excess dry air. (a) What is the molar composition of the combustion products? (b) What is the dew point of the combustion products if the pressure is 14.7 psia?

7-6 A company is considering raising the reaction temperature of two continuously stirred tank reactors in series. The original temperature is 30 °C, and it will be raised to 70 °C. The rate constant is:

$$k = 1230^{-6000/T}\ \text{min}^{-1}$$

The feed is 1.5 weight% acetic anhydride, and the product is acetic acid via a first-order reaction. The feed rate is 100 gpm, and the reactors are 1000 gallons each. (a) How much increase in throughput could be obtained by raising the temperature? (b) What other considerations will influence the final decision to increase the temperature?

7-7 A tank containing 500 gallons of 10 weight% sulfuric acid at 100 °F is fed 40 gpm of a solution of 15 weight% sulfuric acid at 180 °F. The capacity of the tank is 1000 gallons. What are the temperature and concentration of the solution in the tank after 30 minutes?

7-8 A company is considering using two single-stage pumps in place of its present double-stage pumps. To maintain low downtime, three double-stage pumps are operated in parallel. Downtime costs $1000 per hour, and the system operates 365 days per year. The following cost data are available:

	single-stage	double-stage
capital cost	$12,500	$20,000
reliability	75%	50%

What is the optimum arrangement of pumps to minimize cost?

7-9 What is the rate of condensate formed in a 100 foot long horizontal section of 6″ schedule-40 uninsulated iron pipe carrying dead-headed 30 psig steam? The pipe is located in a building, and the condensate is constantly removed by traps.

7-10 An investment company is considering the purchase of an apartment building for $500,000. Depreciation is straight line over 15 years. Operating cost is $25,000 per year. Revenue from rent is $40,000 per year for the first three years and $70,000 per year thereafter. The property is sold at the end of the seventh year for $675,000. Taxes are 40%. What is the rate of return after taxes?

Chemical Engineering Professional License Sample Exam #7 B

7-11 What is the pH of a 100,000 liter pond, originally neutral, after adding 10 kgmole of phosphoric acid (H_3PO_4) and 5 kgmole of NaOH? The values of the dissociation constants for phosphoric acid are:

$$k_1 = 5.9 \times 10^{-3}$$
$$k_2 = 6.15 \times 10^{-8}$$
$$k_3 = 4.8 \times 10^{-13}$$

7-12 Phosphine gas decomposes at 1200 °F from 4.6 atm according to the reaction:

$$4\,PH_3 \longrightarrow P_{4\,(g)} + 6\,H_2$$

The reaction proceeds within first-order rate constant $k = 10\ \text{hr}^{-1}$. What size plug flow reactor (PFR) is needed to obtain 80% conversion with a feed containing 4 lbmole/hr phosphine and 1 lbmole/hr inerts?

7-13 A mixture of polluted air and reaction vent gas is treated in a scrubber according to the diagram shown. The polluted air at 80 °F, 50% relative humidity, consists of 180 lbmoles/hr of dry air, and the reaction vent gas consists of 25.9 lbmoles/hr H_2O and 5.18 lbmoles/hr air. If the gases exit at 72 °F and consist of 5 lbmole/hr of H_2O and 185 lbmole/hr air with a 5 gpm feed and 5.86 gpm bleed on the 30 gpm recirculation stream, find the cooling duty and flow rate of the cooler and the temperature of the scrubber bottoms.

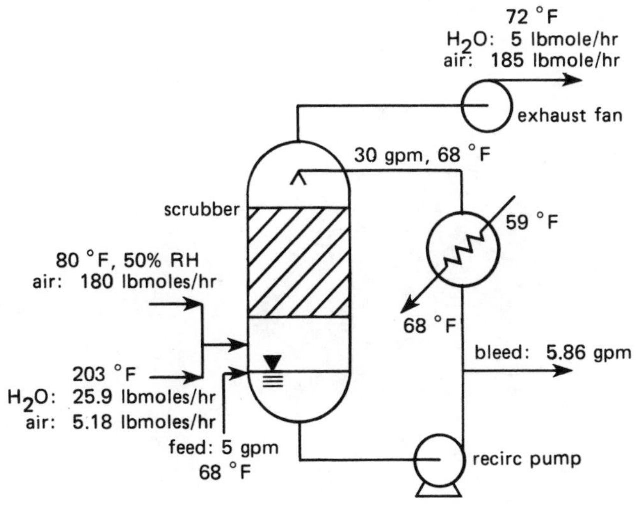

7-14 The system shown is used to heat a feed stream and cool the product from a reactor. The feed rate is 10 kg/s at 40 °C, and the feed and product streams of the reactor are at 250 °C. The carbon steel heat exchangers cost \$200/m². The heat transfer coefficients of the exchangers are:

$$U_I = 500\ \text{J/m}^2\text{-s-°C}$$
$$U_H = 850\ \text{J/m}^2\text{-s-°C}$$

Saturated steam is available at 260 °C at a cost of \$3.00/GJ. The heat capacity, c_p, of the process fluid is 2 kJ/kgmole-°C. The total capital cost is four times the heat exchanger cost, and the annual cost of capital is 25% of the total capital cost. What is the total annual operating and capital cost of the heat exchanger system if the interchanger feed stream leaves at 170 °C?

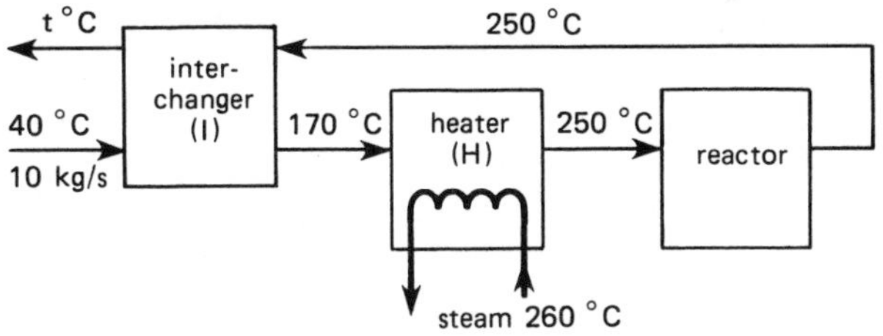

7-15 How much saturated steam at 1000 psia is required to produce one megawatt of power through a turbine-generator exhausting at 3 psia? Assume that the turbine is 80% efficient and the generator is 90% efficient.

7-16 A one shell pass, two tube pass heat exchanger cools 200,000 lbm/hr of water from 100 °F to 90 °F, using river water that enters the shell side at 75 °F and exits at 90 °F. The shell side heat transfer coefficient is 1000 BTU/hr-ft²-°F. The tubes are made of No. 18 BWG copper tubing, each 16 feet long. (a) With a maximum pressure drop of only 10 psi across the tube side fluid, how many tubes are required? (b) What diameter tubes should be used?

7-17 What is the steady state temperature of a galvanized iron roof insulated from a room below on a clear night with the air temperature at 80 °F? Assume that the night sky is a blackbody radiator at −40 °F and the convection heat transfer coefficient between the air and roof is 5 BTU/hr-ft²-°F.

7-18 A double-effect evaporator is used to concentrate 10,000 lbm/hr of a 10% sugar solution to 30%. The feed enters the second effect at 70 °F, and saturated steam enters the first effect at 230 °F. The temperature in the final condenser is 100 °F. The overall heat transfer coefficients in the evaporators are 400 and 300 BTU/hr-ft^2-°F for the first and second effects, respectively. The heating areas for the two effects are the same, and the specific heat is constant at 0.95 BTU/lbm-°F. What is the temperature in each effect, the heating surface for each effect, and the steam consumption in lbm/hr?

7-19 A square-edged orifice is needed to measure the flow rate of air at 50 psig and 70 °F in a 6″ schedule-40 pipe. What size orifice is needed if the maximum air flow is 1000 cfm and the maximum instrument pressure reading across the orifice is 100″ H_2O?

7-20 Copper ore containing $CuSO_4$ is to be extracted in a countercurrent stage extractor. A charge consisting of 10 tons of gangue, 1.2 tons of $CuSO_4$, and 0.5 tons of water is to be treated each hour. The strong solution produced is 10% $CuSO_4$ by weight, and the inert gangue contains 2 tons water/ton gangue (including the $CuSO_4$ dissolved in the water). How many stages are required if the recovery of $CuSO_4$ is to be 98% of that in the ore?

1-1 Assumptions:

1. No pumps for 1st 10 years
2. TANKS have infinite life & one time cost
3. No inflation. Salvage = $0

Cash flow: AS STATED

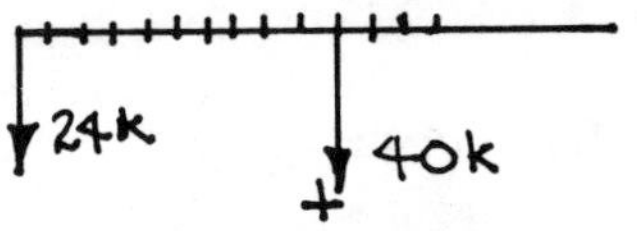

one time costs

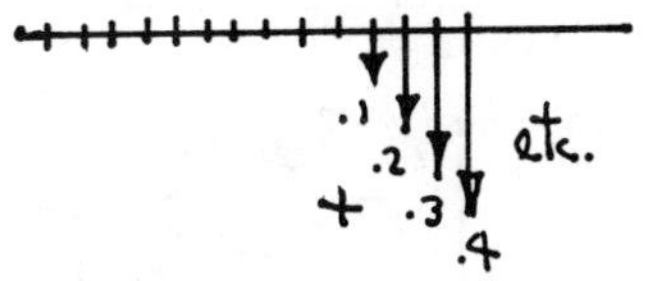

GRADIENT MAINT. COST

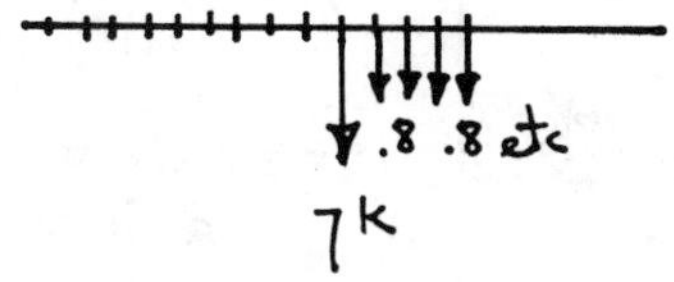

PUMP cost + FIXED MAINT

PROCEDURE: Find PW of one time costs, FIND EUAC of recurring costs
MAKE NEW CASH FLOW

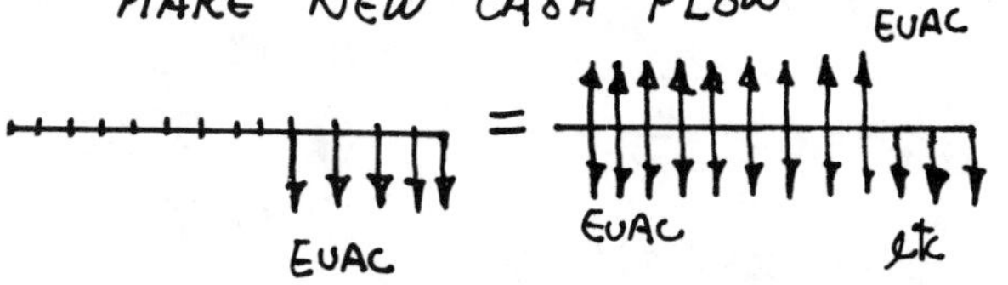

then $\text{CAP COST} = PW(\text{one time}) + \dfrac{EUAC}{i}$

$\text{CAP COST} = 24 + 40[P/F, 10\%, 10] - EUAC(P/A, 10\%, 10) + \dfrac{EUAC}{i}$

$EUAC = 0.8 + 0.1[P/G, 10\%, 7][A/P, 10\%, 7] + 7[A/P, 10\%, 7]$

$EUAC = 0.8 + 0.1[12.7631][0.2054] + 7[.2054]$

$= \$2.500\text{ k/year}$

$\text{CAP COST} = 24 + 40[0.3855] - 2.500[6.1446] + 2.500/0.10$

$\text{CAP COST} = \$49.059^{K}$ or $\$49{,}059$

1-2

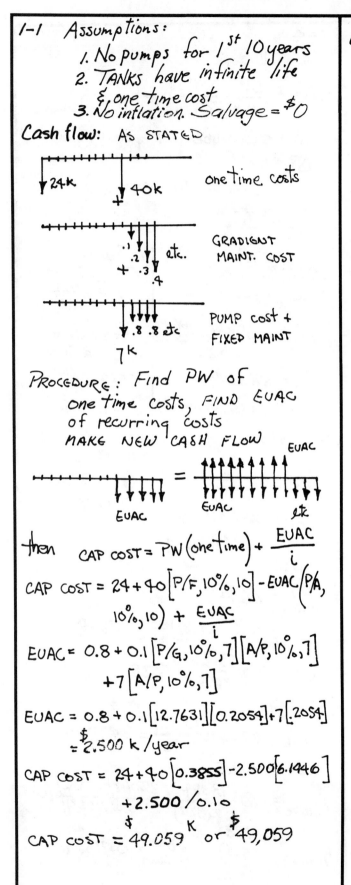

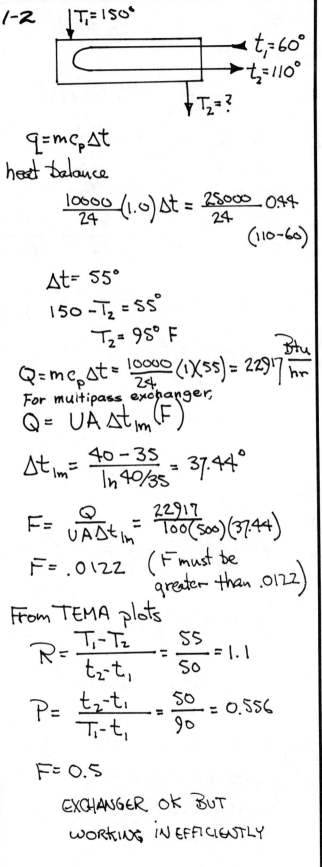

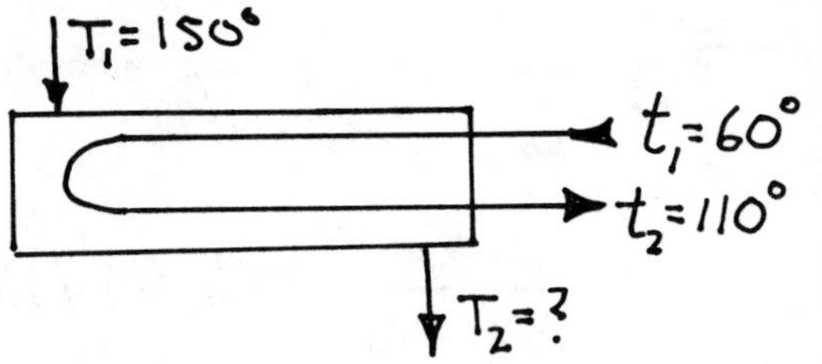

$q = mc_p\Delta t$

heat balance

$$\frac{10000}{24}(1.0)\Delta t = \frac{25000}{24}\,0.44\,(110-60)$$

$\Delta t = 55°$

$150 - T_2 = 55°$

$T_2 = 95°\text{ F}$

$$Q = mc_p\Delta t = \frac{10000}{24}(1)(55) = 22917\ \frac{\text{Btu}}{\text{hr}}$$

For multipass exchanger,

$Q = UA\,\Delta t_{lm}(F)$

$$\Delta t_{lm} = \frac{40-35}{\ln 40/35} = 37.44°$$

$$F = \frac{Q}{UA\,\Delta t_{lm}} = \frac{22917}{100(500)(37.44)}$$

$F = .0122$ (F must be greater than .0122)

From TEMA plots

$$R = \frac{T_1 - T_2}{t_2 - t_1} = \frac{55}{50} = 1.1$$

$$P = \frac{t_2 - t_1}{T_1 - t_1} = \frac{50}{90} = 0.556$$

$F = 0.5$

EXCHANGER OK BUT WORKING INEFFICIENTLY

1-3

2nd order irreversible

$$\ln \frac{M - X_A}{M(1-X_A)} = C_{A_0}(M-1)kt$$

[a] by decreasing temperature keeping M, X_A, C_{A_0} constant increases k all other terms cancel

at 100°C: $k_1 t_1 = \frac{1}{C_{A_0}(M-1)} \ln \frac{M-X_A}{M(1-X_A)}$

at 120°C: $k_2 t_2 = \frac{1}{C_{A_0}(M-1)} \ln \frac{M-X_A}{M(1-X_A)}$

$\therefore \quad k_1 t_1 = k_2 t_2$

$17.4 t_1 = 30 t_2$

$t_2 = 0.58 t_1$

or time is reduced by 58% of t_1 by raising temperature to 120°C

[b] at 100°C $M = 4.98$ (let $X_A = .5$ for example)

$$\ln \frac{4.98 - 0.5}{4.98(0.5)} = 0.58734 = C_{A_0}(4.98-1)k_1 t_1$$

$M = 4.0$

$$\ln \frac{4-.5}{4(.5)} = .5596 = C_{A_0}(4-1)k_2 t_2$$

$t_2 = 1.264 t_1$ since $k_1 = k_2$ in this case

time is increased 1.264 times by decreasing M to 4.0.

$\therefore$ increasing temperature to 120°C results in lesser reaction time.

1-4

Bernoulli

$$\Delta z + \frac{\Delta P}{\rho} + \frac{\Delta v^2}{2g} + h_f + h_{pump} + h_{valve} = 0$$

$\Delta P = 0$ (scrubber pot operating at 1 atm)

$\Delta z = 60$ ft

$v_1 \cong 0$ (negligible velocity)

$$\therefore \quad 60 + \frac{v_2^2}{2g} + h_f + h_{pump} + h_{valve} = 0$$

Crane:

$$C_V = Q\sqrt{\frac{sp.g.}{\Delta P_{VALVE}}}$$

$v_2 = 4.78$ ft/sec @ 50 gpm

$v_2 = 4.78\left(\frac{140}{50}\right) = 13.38$ ft/sec @ 140 gpm

$\mu = .86$ $\rho = 62.2$

sp.g. = .9968 @ 80°F

50 gpm.

$$N_{Re} = 123.9 \frac{dv\rho}{\mu}$$

$$N_{Re} = 123.9 \frac{(2.067)(4.78)(62.2)}{0.86}$$

$N_{Re} = 88538$

$\varepsilon/D = .0002$ (assumed)

$f = .019$

$$h_{valve} = -60 - \frac{4.78^2}{2(32.2)} - .019\frac{220(12)(4.78)^2}{(2.067)64.4} + 135$$

$h_{valve} = 66$ ft

$$\Delta P_{valve} = 66(.4331)\frac{62.2}{62.4} = 28.49 \text{ psia}$$

$$C_V = 50\sqrt{\frac{.9968}{28.49}} = 9.35$$

140 gpm

$$N_{Re} = 88538\frac{13.384}{4.78} = 247906$$

$f = 0.0165$

$$h_{valve} = -60 - \frac{13.384^2}{2(32.2)} - \frac{220(12)(13.384^2)}{2.067(64.4)}(.0165) + 123$$

$h_{valve} = 1.64$ ft $\quad \Delta P_{valve} = 0.71$ psia

$$C_V = 140\sqrt{\frac{.9968}{.71}} = 166.$$

1–5 $\frac{1}{2}N_2 + \frac{1}{2}O_2 \rightleftharpoons NO$

$$K = 2\times10^{-3} = \frac{[NO]}{[N_2]^{1/2}\,[O_2]^{1/2}}$$

assume: 1 mole air at start

X = moles NO

$0.79 - \frac{X}{2}$ = moles N_2

$0.21 - \frac{X}{2}$ = moles O_2

$$\therefore\ 2\times10^{-3} = \frac{X}{\left(0.79-\frac{X}{2}\right)^{1/2}\left(0.21-\frac{X}{2}\right)^{1/2}}$$

$$\frac{.1659}{X^2} - \frac{1}{2X} + \frac{1}{4} = 2.5\times10^{5}$$

since 1/4 is small compared to 2.5×10^5

$$\frac{1}{X}\left(\frac{.1659}{X} - \frac{1}{2}\right) = 2.5\times10^{5}$$

if X is small so that $\frac{.1659}{X} \gg \frac{1}{2}$

then $X \cong 0.0008146$

(by the bisection method $X = 0.00081362$)

mole fraction NO:

$$\frac{0.0008146}{\left[.79-\frac{.0008146}{2}\right]+\left[.21-\frac{.0008146}{2}\right]} = 0.0008146 \text{ or } 0.08146\%$$

1–6 refer to McCabe-Thiele plot next page

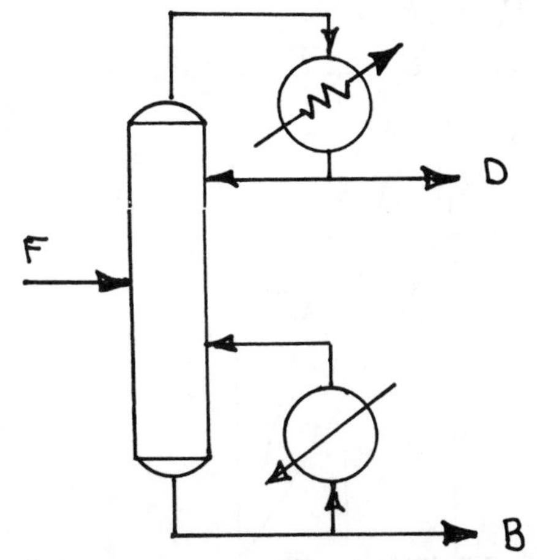

Assume:

1. Tower acts under pressure so feed is liquid
2. Feed is 100 kg/hr

1–6 (con't)

let: W_A = weight fraction benzene

M_A = mole wgt benzene = 78

M_B = mole wgt toluene = 92

$M_A/M_B = 0.8478$

$$X_A = \frac{1}{1 + \frac{M_A}{M_B}\left(\frac{1-W_A}{W_A}\right)}$$

MATERIAL BALANCE:

$D = (.9)(.6)(100)/.924 = 58.44$ kg/hr

$B = 100 - 58.44 = 41.56$ kg/hr

$(X_A)_{feed} = 1/(1 + .8478(.4/.6)) = .639$

$(X_A)_D = 1/(1 + .8478(.076/.924)) = .935$

$(W_A)_B = (0.1)(60)/41.56 = 0.144$

$(X_A)_B = 1/(1 + .8478(.856/.144)) = .1656$

from graphical construction

$$\frac{R_m}{R_m+1} = \frac{.935 - .572}{.935 - 0} = .3882$$

$R_m = .6345$; $R_{ACTUAL} = (.6345)1.3 = .8249$

ACTUAL rectifying section slope

$$R/R+1 = .8249/1.8249 = .4520$$

intercept of rect. line @ $X = 0$:

$$-.4520(.935) + .935 = 0.5124$$

from construction:

theoretical stages:

$n_T = 9+ \Rightarrow$ use $n = 10$

$n_{actual} = 10/0.6 = 16.67$

height = n_{actual} (HETP) = 16.67(.2)

= 3.33 meters

PROBLEM 1-6

Solution

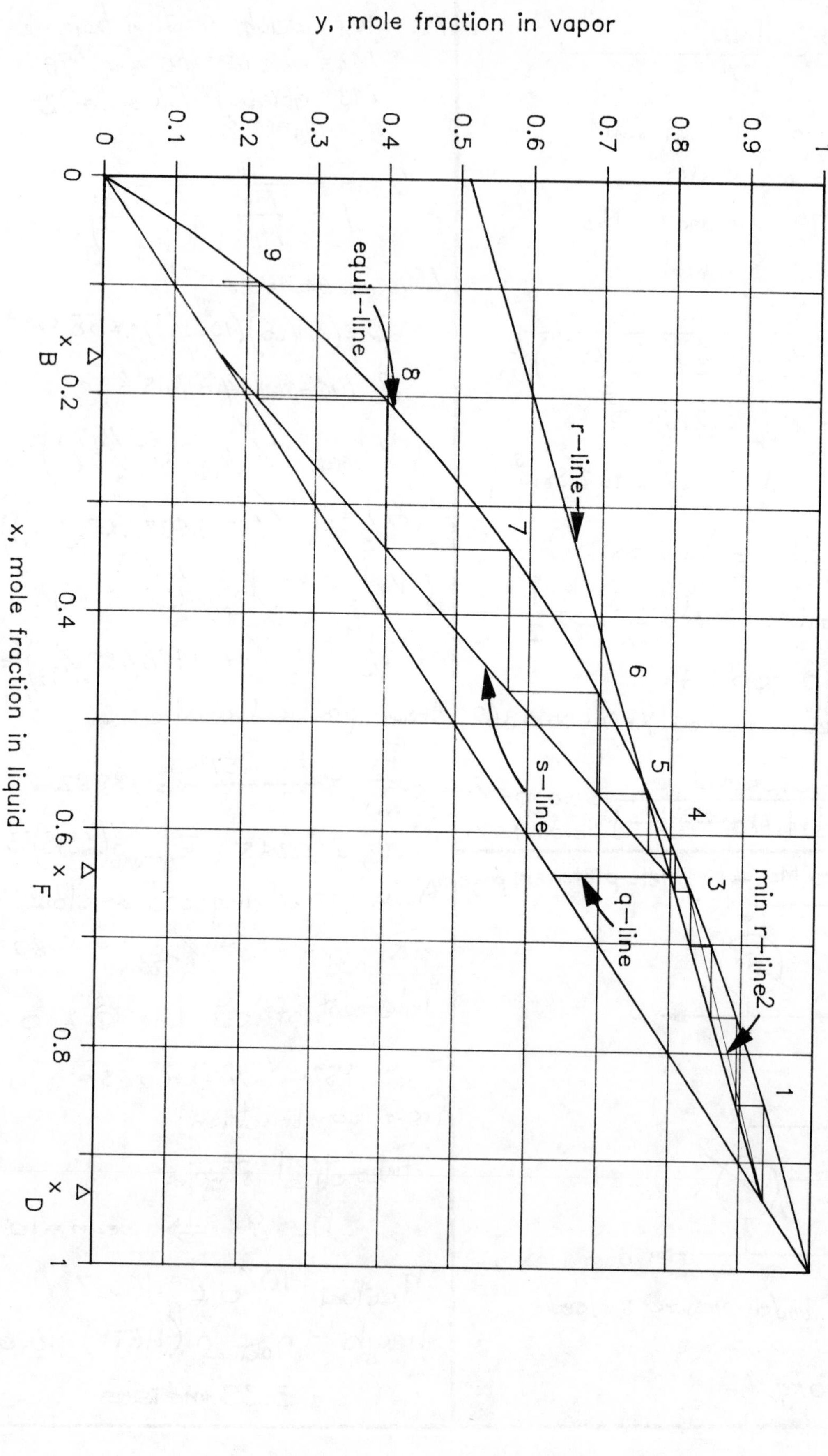

1-7

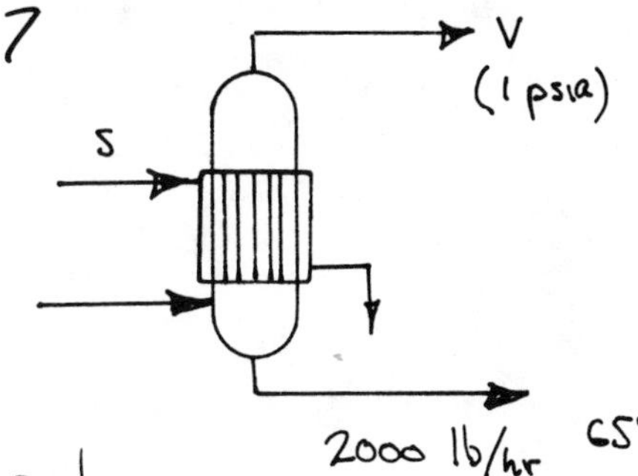

Mat Balance

$F = V + 2000$

$0.05F = 0.65(2000)$

$F = 26000$ lb/hr

$V = 24000$ lb/hr

neglect B.P.E.

assume C_p NaOH = 1.0 (dilute)

[a] heat bal: $Q = 26000(C_p)(t_{evap} - t_{feed}) + 24000\,\lambda_{evap}$

steam tables:

$t_{evap} = 102°F \quad \lambda_{evap} = 1036 \frac{Btu}{lb}$

$Q = 26000(1)(102-70) + 24000(1036)$

$Q = 2.5696 \times 10^7 \frac{Btu}{hr}$

$Q = UA\Delta t = 2.5696 \times 10^7$

$$\Delta t = \frac{2.5696 \times 10^7}{(300)(2500)}$$

$\Delta t = 34.3$ °F

Evaporator OK since 5 psig steam $t_{sat} = 226°F$

[b] use 5 psig steam since it has highest latent heat ($960 \frac{Btu}{lb}$) versus 919 & 880 for 40 and 100 psig, respectively

[c] control pressure in steam chest (since this controls temp.)

1-8

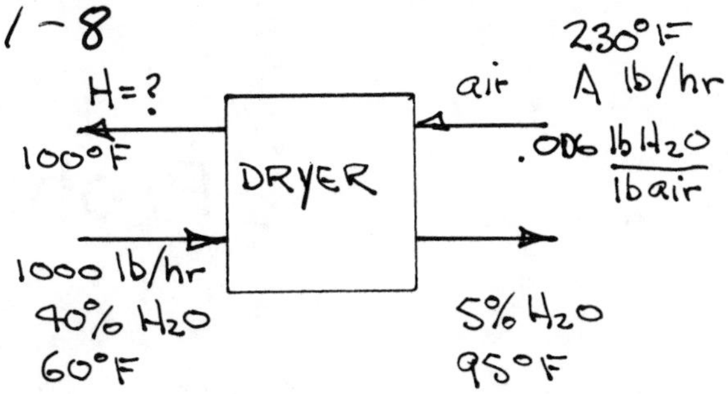

material balance

dry material: $0.6(1000) = 600$ lb/hr

exit material: $600/.95 = 631.6$ lb/hr

H_2O in exit: $631.6 - 600 = 31.6$ lb/hr

H_2O evap to air: $.4(1000) - 31.6 = 368.4 \frac{lb}{hr}$

heat balance $\quad \Delta h = mc_p\Delta t$

$\Delta h_{dry\,air} + \Delta h_{moisture} =$

$\Delta h_{dry} + \Delta h_{water} + \Delta h_{evap} + \Delta h_{water\,vapor}$

$A(.231)(230-100) + .006A(.48)(230-100) =$

$600(.2)(95-60) + 400(1)(68-60)$

$+ 368.4(1052) + 368.4(.48)(100-68)$

$+ 31.6(1)(95-68)$

$A = 13204.3$ lb dry air/hr

incoming air:

$1.006A = 13283.5 \frac{lb}{hr}$ air @ $H = 0.006 \frac{lb\,H_2O}{lb\,air}$

$$H_{exit} = \frac{.006A + 368.4}{A} = .0339 \frac{lb\,H_2O}{lb\,dry\,air}$$

1-9

see graphical solution next page

Problem 1-9

E1:	R1
514 lb	186 lb
52% C	R2:
12% A	93% A
36% B	5.5% B
	1.5% C

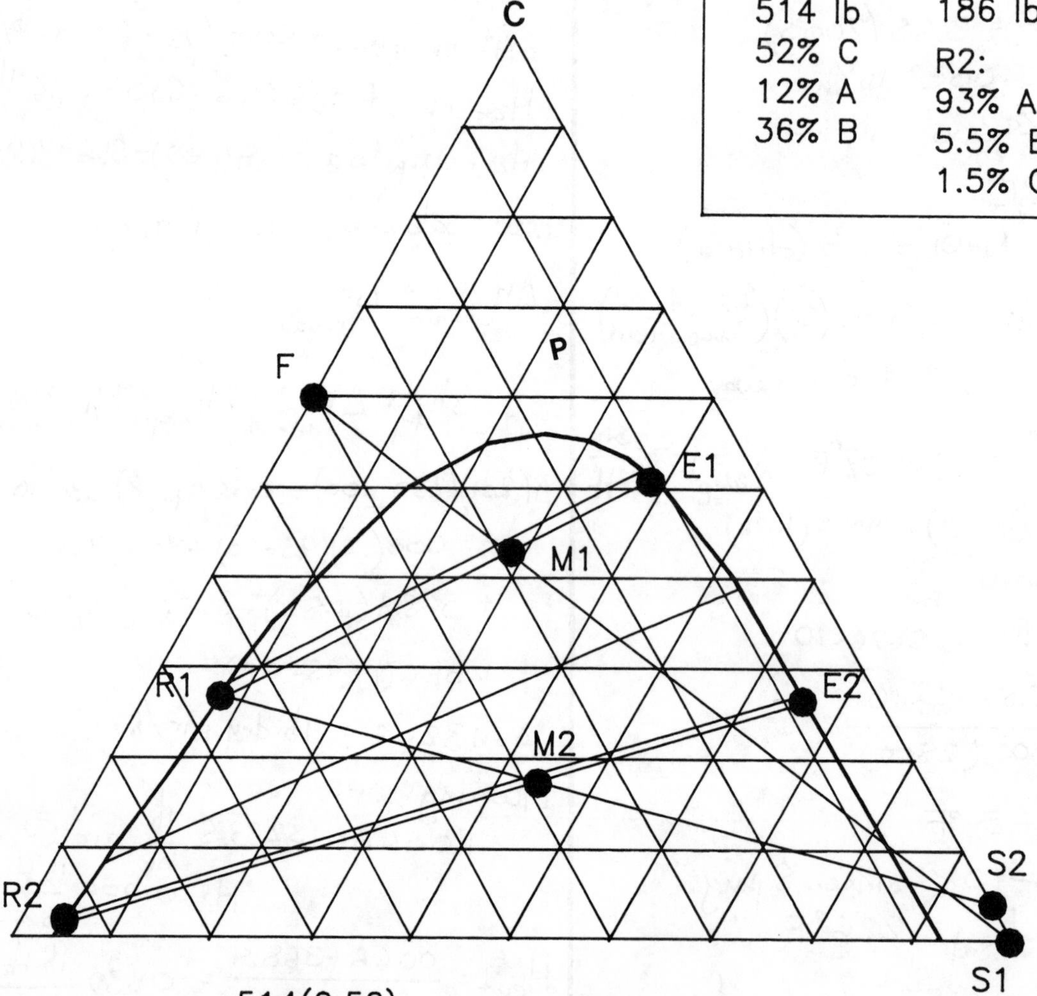

$$\% \text{ extr} = \frac{514(0.52)}{500(0.60)} \times 100 = 89\%$$

1-10

Assumptions:
1. constant density $\varepsilon_A = 0$
2. elementary reaction
3. 2nd order, irreversible

2nd order CSTR: $k\tau = \dfrac{C_{A_0} - C_A}{C_A C_B}$

A = NaOH
B = EtAc

$C_{A_0} = .00587(1.6)/2.8 = 0.003354$

$C_{B_0} = 0.0389(1.2)/2.8 = 0.01667$

$C_B = C_{B_0} - C_{A_0} + C_A = 0.01667 - 0.003354 + C_A$

$C_B = 0.013316 + C_A$

25°C

$\tau = 0.602/(1.6+1.2) = 0.215\ \text{hr.}$

$C_A = .001094$

$C_B = .013316 + .001094 = .01441$

$k = \dfrac{.003354 - .001094}{(.001094)(.01441)(.215)} = 666.8\ \dfrac{l}{\text{mole·hr}}$

40°C

$C_A = .0009243$

$C_B = .013316 + .0009243 = .01424$

$k = \dfrac{0.003354 - 0.0009243}{(0.0009243)(0.01424)(.215)} = 858.6\ \dfrac{l}{\text{mole·hr}}$

since $\ln \dfrac{k_2}{k_1} = \dfrac{E}{R}\left[\dfrac{1}{T_1} - \dfrac{1}{T_2}\right]$

$\ln \dfrac{858.6}{666.8} = 0.253 = \dfrac{E}{R}\left[\dfrac{1}{298} - \dfrac{1}{313}\right]$

$\dfrac{E}{R} = 1573.2 \qquad E = 1573.2(1.987) = 3126\ \text{cal/mole}$

$k = k_0 e^{-E/RT}$

$k_0 = 666.8\, e^{1573.2/298} = 1.308 \times 10^5$

$k\tau = \tau k_0 e^{-E/RT} = \dfrac{C_{A_0} - C_A}{C_A C_B}$

$(.215)(1.308 \times 10^5) e^{-1573/T} = \dfrac{C_{A_0} - C_A}{C_A(0.013316 + C_A)}$

1-11

C_p° at 140°F (ideal → zero pressure)

140°F → 333°K

$C_p^\circ = 6.8214 + 5.7095 \times 10^{-3}(333) - 5.107 \times 10^{-6}(333^2) + 1.547 \times 10^{-9}(333)^3$

$C_p^\circ = 8.213\ \text{cal/gmole °K}$

Critical properties of Cl_2

$T_c = 144°C = 417°K$

$P_c = 76.1\ \text{atm}$

$T_R = 333/417 = 0.8$

$P_R = 112/(14.7 \times 76.1) = 0.10$

from generalized C_p charts at $T_R = 0.8$, $P_R = 0.1$

$C_p - C_p^\circ = 1.70$

$\therefore\ C_p = 1.70 + 8.214 = 9.914\ \text{cal/gmole °K}$

1-12

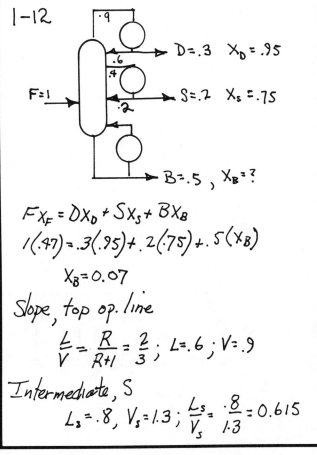

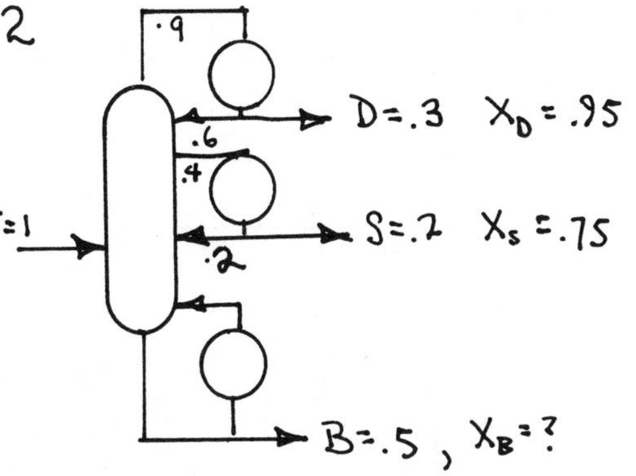

$F x_F = D x_D + S x_s + B x_B$

$1(.47) = .3(.95) + .2(.75) + .5(x_B)$

$x_B = 0.07$

Slope, top op. line

$\dfrac{L}{V} = \dfrac{R}{R+1} = \dfrac{2}{3};\quad L = .6;\ V = .9$

Intermediate, S

$L_s = .8,\ V_s = 1.3;\quad \dfrac{L_s}{V_s} = \dfrac{.8}{1.3} = 0.615$

PROBLEM 1-12

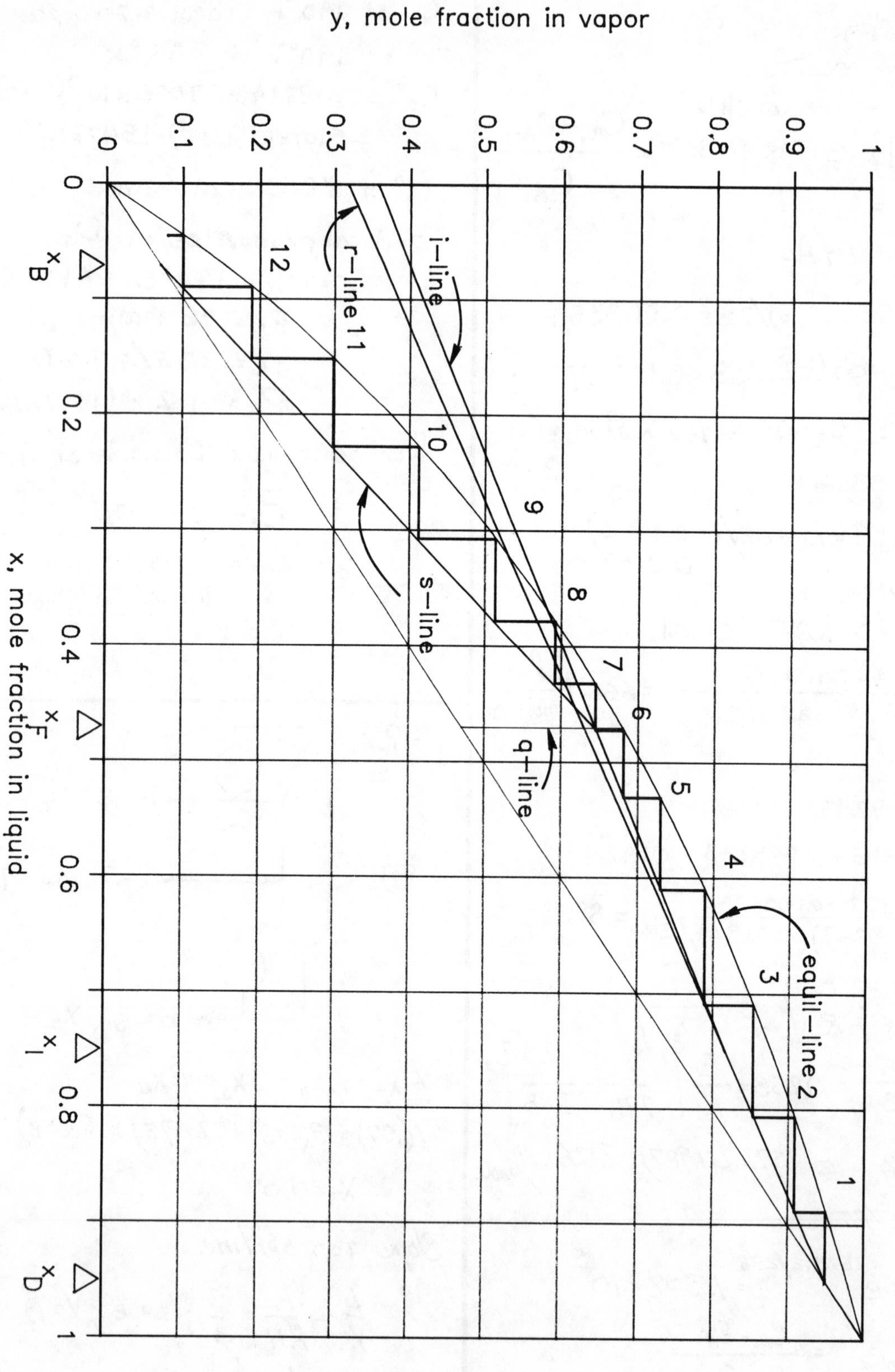
y, mole fraction in vapor
x, mole fraction in liquid
0
0.1
0.2
0.3
0.4
0.5
0.6
0.7
0.8
0.9
1
0
0.2
0.4
0.6
0.8
1
x_B
x_F
x_1
x_D
i-line
r-line
s-line
q-line
equil-line
1
2
3
4
5
6
7
8
9
10
11
12

1-12 (cont.)

Assume Feed & side stream enter as sat liquid.

FROM PLOT:

[A] theoretical plates = 9+ not including reboiler

[B] feed plate: 6

[C] side stream plate 3

1-13

$V = 10000$ gal; t = thickness, in

h = height, ft P = pressure, psi

D = diameter, ft

d = diameter inches

$V = \pi D^2 h/4$; $h = 4V/\pi D^2 = 1702/D^2$

C = cost \$ = top \$ + bottom \$ + side \$

$$C = 4\frac{\pi D^2}{4} + 9\frac{\pi D^2}{4} + 21t\pi D h$$

$$C = 10.21D^2 + 21053.8/D$$

min $\frac{dC}{dD} = 0 = 2(10.21)D - 21053.8/D^2$

$D^3 = 1031.04$

$D = 10.103$ ft $h = \frac{1702}{D^2} = 16.68$ ft

check min thickness

$$P = \frac{(62.4)(1.525)h}{144} = 11.02 \text{ psi}$$

$t_{min} = \frac{Pd}{2400} = 0.0557$; OK: $t_{min} < .1875$

use .1875

$C = 10.21(10.103)^2 + 21053.8/10.103$

$C = \$3126.06$

1-14

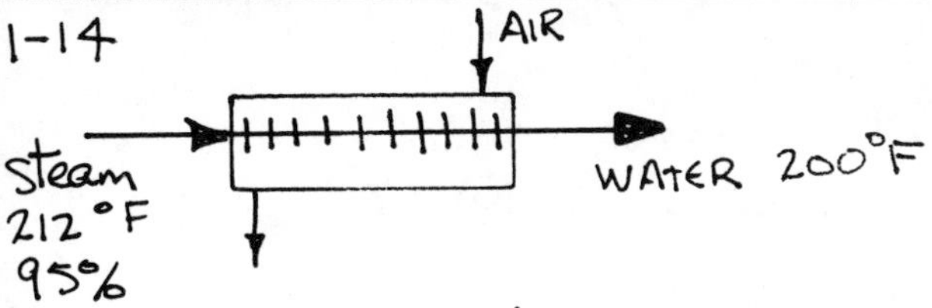

(a) w = lb steam/hr

$w((.95h_v) + .05h_L - h_L - C_p(212-200)) = 60000$

$w(.95(1150.8) + .05(181.11) - 181.11 - (1)(12)) = 60000$

$w = 65.99$ lb/hr

b) $Q = UA\Delta t$ (t = air temp rise! cannot use Δt_{lm} because of condensing phase)

$A = [.5 + 2.5(.8)]L = 2.5L$

$$U = \frac{1}{\frac{1}{h_s} + \frac{1}{h_a}} = \frac{1}{\frac{1}{200} + \frac{1}{5}} = 4.878$$

$60000 = 4.878(2.5L)(30)$; $L = 164$ ft

1-15

$$K\tau = \frac{C_{A_0} - C_A}{C_A C_B}; \quad \tau = \frac{2}{8} = .25 \text{ min}$$

find K: $C_A = C_{A_0}(1 - X_A) = 2(1 - .6) = .8$

$C_B = 3 - (C_{A_0} - C_A) = 1 + C_A = 1.8$

$$K = \frac{1}{.25}\,\frac{2 - 0.8}{0.8(1.8)} = 3.333 \frac{\text{ft}^3}{\text{lbmole-min}}$$

[a] for 2nd reactor

$C'_{A_0} = 0.8$ $C'_{B_0} = 1.8$ $C'_B = 1 + C'_A$

$$K\tau = .8333 = \frac{C'_{A_0} - C'_A}{C'_A(1 + C'_A)} = \frac{.8 - C'_A}{C'_A(1 + C'_A)}$$

$C'_A = 0.3731$

$$X_{A\,overall} = 1 - \frac{C'_A}{C_{A_0}} = 1 - \frac{.3731}{2} = .8135$$

[b] $C'_A = .3731$ $C'_B = 1.3731$ $C_R = C_S = 1.6269$

$C_R = C_S = C_{A_0} - C'_A = 1.6260$

1-16

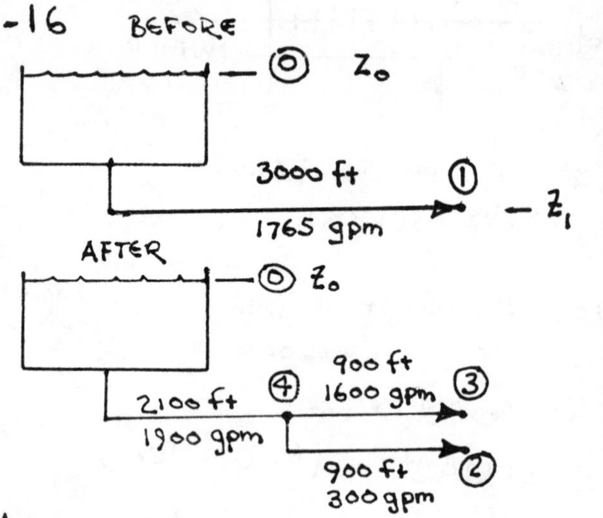

Assume:

1. $z_1 = z_2 = z_3$
2. Same pressure at 0 & 1,2,&3
3. pond large so $v_0 \cong 0$

LET h_{f1} = friction loss from 0→1

h_{f2} = friction loss 4→2

h_{f3} = friction loss 4→3

h_{f4} = friction loss 0→4

The overall Bernoulli equation

$$\Delta z + \frac{\Delta v^2}{2g} + \frac{\Delta P}{\rho} + h_f = 0$$

Reduces to

$$z_0 = h_{f1} + \frac{v_1^2}{2g} \quad \text{before}$$

$$z_0 = h_{f4} + h_{f3} + h_{f2} + \frac{v_2^2}{2g} + \frac{v_3^2}{2g} \quad \text{after}$$

z_0 must be equal or less than the before condition

given:

Point	L (ft)	Q (gpm)	z	D (ft)	d (in)
1	3000	1765	0	.665	7.981
2	900	300	0	.665	7.981
3	900	1600	0	.665	7.981
4	2100	1900	0	.665	7.981

constants

$\rho = 62.34 \frac{lb}{ft^3}$; $\mu = 1.13$ cp; $g = 32.2 \frac{ft}{sec^2}$

$$v = 0.408 \frac{Q}{d^2}$$

$$N_{Re} = 50.6 \frac{Q\rho}{d\mu}$$

$$f = 0.0035 + 0.027 [N_{Re}]^{-.42}$$

breaking down the reduced Bernoulli equations noting that

$$h_f = \frac{fL}{D} \frac{v^2}{2g}$$

before:

$$z_0 = \frac{\left[1 + \frac{f_1 L_1}{D_1}\right] v_1^2}{2g}$$

after:

$$z_0 = \frac{\left[\frac{f_4 L_4 v_4^2}{D_4}\right] + \left[1 + \frac{f_2 L_2}{D_2}\right] v_2^2 + \left[1 + \frac{f_3 L_3}{D_3}\right] v_3^2}{2g}$$

Point	v	N_{Re}	f
1	11.31	617343	0.0036
2	1.92	109931	0.0037
3	10.25	559631	0.0036
4	12.17	664561	0.0036

z_0(before) = 34.21 ft

z_0(after) = 36.05 ft

To achieve flow requirements, the elevation of the pond has to increase by 1.84 feet. The req'd flow therefore cannot be obtained as stated.

1-17

(a) $CO + 3H_2 \rightleftharpoons CH_4 + H_2O$

Given:

component	gef (cal/gmole°K)	H_f (cal/gmole)	H (cal/gmole)
CO	-47.254	-27020	5700
H_2	-31.186	0	5537.4
CH_4	-45.21	-15987	8321.
H_2O	-45.128	-57107	6689.6

T = 800 °K

R = 1.987 cal/gmole°K

$$G = T\,gef + H_f + H$$

$$\Delta G = G_{CH_4} + G_{H_2O} - \left[G_{CO} + 3G_{H_2}\right]$$

i	G_i cal/gmole
CO	-59123.2
H_2	-19411.4
CH_4	-43834
H_2O	-86519.8

$\Delta G = -12996.4$ cal/gmole

since

$$\Delta G = -RT \ln K$$

$$K = \exp\left[\frac{-\Delta G}{RT}\right]$$

K = 3554.

let x = moles CO at equilibrium

start with: 1 mole CO

3 moles H_2

total moles = 4-2x

$$K = \frac{(4-2x)^2 x^2}{(1-x)(3-3x)^3}$$

trial & error

x	K
0.9196	3499.77
0.9197	3517.35
0.9198	3535.04
0.9199	3552.84 ←
0.92	3570.75

x = 0.9199

conversion = x = 0.9199

(b)

$$\Delta H = H_{CH_4} + H_{f_{CH_4}} + H_{H_2O} + H_{f_{H_2O}} - \left[H_{CO} + H_{f_{CO}} + 3\left[H_{H_2} - H_{f_{H_2}}\right]\right]$$

$\Delta H = -53375.6$ cal/gmole

1-18 tower diam calculated @ 75% flood point

$$A = \frac{3600 \text{ lb/hr}}{(.75)(1000) \text{ lb/hr-ft}^2} = 4.8 \text{ ft}^2$$

$$D = \sqrt{\frac{4.8\,(4)}{\pi}} = 2.47 \text{ ft}$$

$$H_{OTG} = \frac{G}{A k_y a (1-y)_{lm}}$$

Since N_2O_4 is dilute, $k_y a \simeq 100 \frac{lbm}{hr\text{-}ft^3}$ & $(1-y) \simeq 1_{lm}$

$$Z = H_{OTG} N_{TOG}$$

$$H_{OTG} = \frac{3600}{100(1)(4.8)}$$

$$= 7.5 \text{ ft}$$

$$N_{TOG} = \frac{y_1 - y_2}{\dfrac{(y_1 - y_1^*) - (y_2 - y_2^*)}{\ln\left[\dfrac{y_1 - y_1^*}{y_2 - y_2^*}\right]}}$$

$$= \ln\frac{y_1}{y_2} = \ln\frac{1/99}{.01/99} = 4.605$$

since N_2O_4 readily reacts

$y^* \sim 0$

$$Z = (0.217)(7.5) = 1.63 \text{ ft}$$

1-19

Antoine equation for vapor press.

$$\log_{10} P = A + \frac{B}{C+t} \qquad P\,[\equiv \text{mm Hg}] \quad t\,[\equiv {}^\circ C]$$

	A	B	C
C_3H_8	6.82973	-813.2	248
C_4H_{10}	6.83029	-945.9	240

at 5 atmospheres:

C_3H_8: $\log_{10}(760(5)) = 6.82973 - \frac{813.2}{248+t}$

$t = 2.2^\circ C$

C_4H_{10}: $\log[760(5)] = 6.83029 - \frac{945.9}{240+t}$

$t = 51^\circ C$

Raoults Law: $(P_i^\circ \equiv \text{atm})$

$$X_{C_3H_8} P^\circ_{C_3H_8} + (1 - X_{C_3H_8}) P^\circ_{C_4H_{10}} = 5$$

$$X_{C_3H_8} = \frac{5 - P^\circ_{C_4H_{10}}}{P^\circ_{C_3H_8} - P^\circ_{C_4H_{10}}}$$

$$y_{C_3H_8} = X_{C_3H_8} P^\circ_{C_3H_8} / 5$$

t	$P^\circ_{C_3H_8}$	$P^\circ_{C_4H_{10}}$	x	y
2.2	5	–	1	1
12.5	6.718	1.597	.6645	.8928
25.0	9.336	2.399	.3749	.7000
37.5	12.606	3.474	.1671	.4213
51.	–	5.0	0	0

1-20
[a]

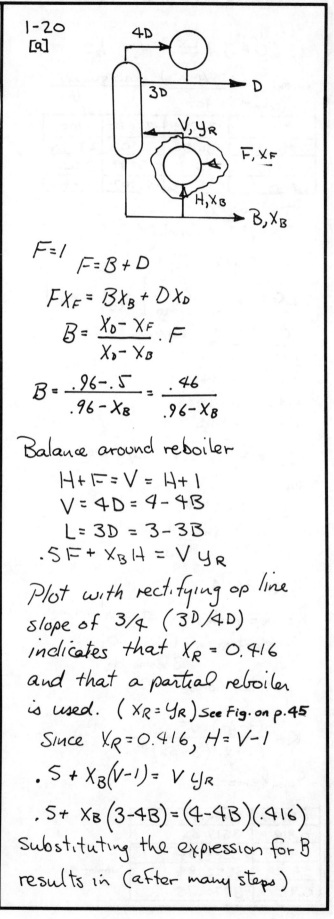

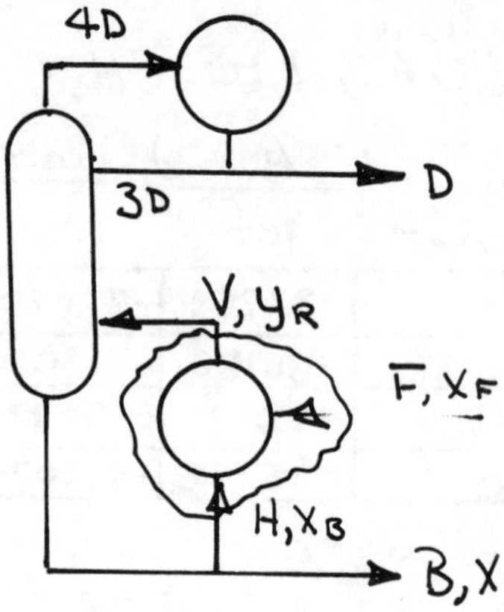

$F = 1 \qquad F = B + D$

$$F x_F = B x_B + D x_D$$

$$B = \frac{x_D - x_F}{x_D - x_B} \cdot F$$

$$B = \frac{.96 - .5}{.96 - x_B} = \frac{.46}{.96 - x_B}$$

Balance around reboiler

$H + F = V = H + 1$

$V = 4D = 4 - 4B$

$L = 3D = 3 - 3B$

$.5F + x_B H = V y_R$

Plot with rectifying op line slope of 3/4 (3D/4D) indicates that $x_R = 0.416$ and that a partial reboiler is used. $(x_R = y_R)$ See Fig. on p.45

Since $x_R = 0.416$, $H = V - 1$

$.5 + x_B(V-1) = V y_R$

$.5 + x_B(3 - 4B) = (4 - 4B)(.416)$

Substituting the expression for B results in (after many steps)

1-20 (cont.)

$$X_B^2 - .73467 X_B + .11733 = 0$$

$$X_B = .23465$$

[b] normal use of
column when feed is
Saturated liquid results
in $X_B = 0.269$ (see plot)

$$.05 \, F = .269 B + .96 D$$

$$F = 1$$

$$D = 1 - B$$

$$.5 = .269 B + .96 - .96 B$$

$$B = .6657 \text{ or } 66.6\% \text{ Bottoms}$$

~~because~~

$$B = \frac{.46}{.96 - X_B} = \frac{.46}{.96 - .269}$$

refer to McCabe-Thiele plots on
following pages

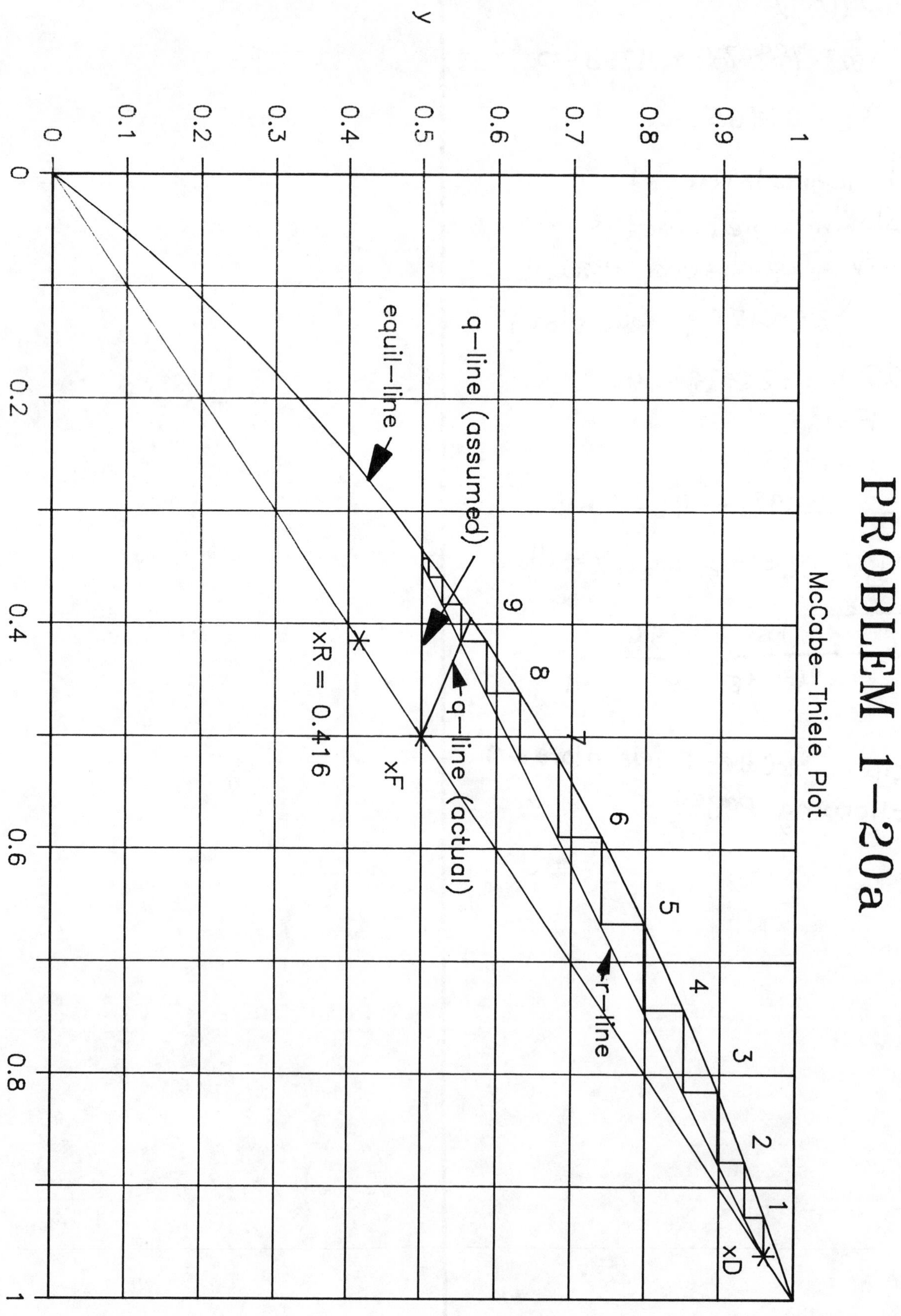
PROBLEM 1-20a
McCabe-Thiele Plot
y
x
q-line (assumed)
equil-line
q-line (actual)
r-line
xR = 0.416
xF
xD
1
2
3
4
5
6
7
8
9
0
0.1
0.2
0.3
0.4
0.5
0.6
0.7
0.8
0.9
1
0
0.2
0.4
0.6
0.8
1

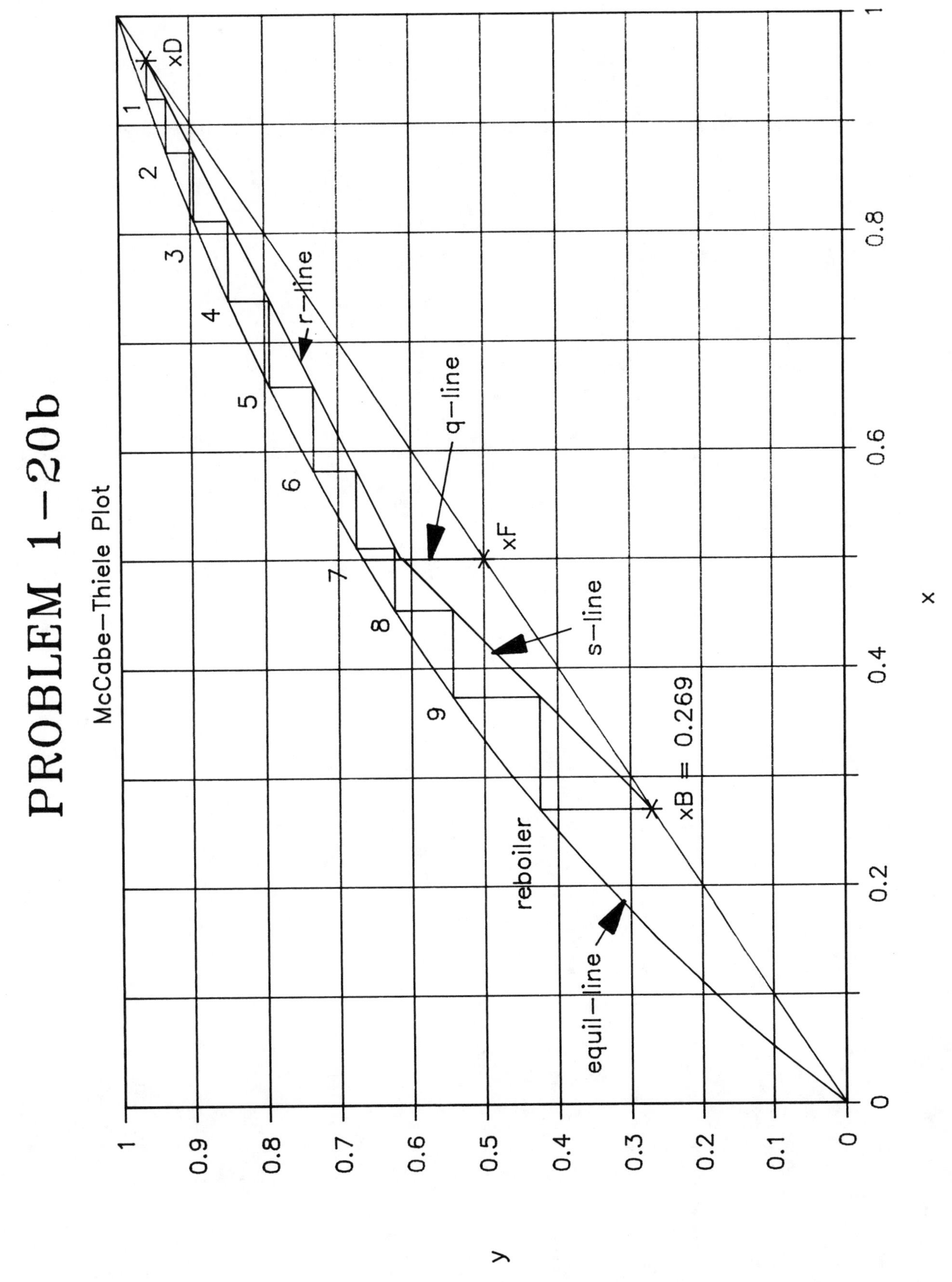
PROBLEM 1 − 20b
McCabe–Thiele Plot
xD
xF
xB = 0.269
r–line
q–line
s–line
equil–line
reboiler
1
2
3
4
5
6
7
8
9
x
y
0
0.2
0.4
0.6
0.8
1
0.1
0.3
0.5
0.7
0.9

2-1 (a)

$PW = -300 - 35(P/A, 8\%, 60) + 50(P/F, 8\%, 60)$

$= -300 - 35(12.3766) + 50(.0099)$

$= -\$732.686K$

(b)

CAP cost = initial + EUAC/i

$EUAC = 35 + (300-50)(A/F, 8\%, 60)$

$EUAC = 35 + 250(.0008) = \$35.2^K$

$CAP\ COST = 300 + 35.2/.08$

$= \$740\ K$

2-2

$G = \dfrac{20 \frac{moles}{hr}\ 29 \frac{lb}{moles}}{\pi(1)^2} = 184.6 \dfrac{lb}{hr\,ft^2}$

$L = \dfrac{(20)(18)}{\pi(1)^2} = 114.6 \dfrac{lb}{hr\,ft^2}$

$H_G = 5.31(184.6)^{0.1}(114.6)^{-.39} = 1.408 ft$

$N_G = Z/H_G = 4/1.408 = 2.84$

(a)

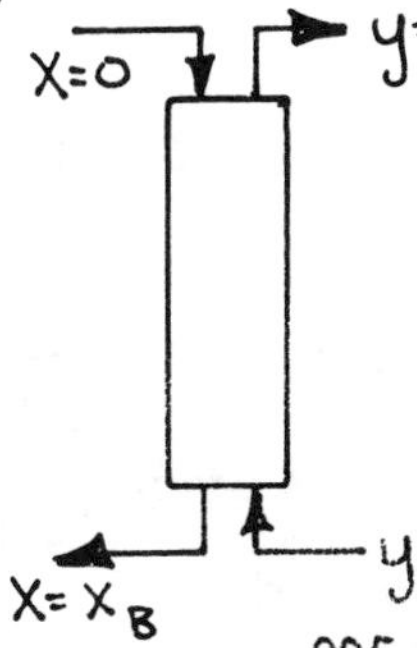

counter flow op line

$y = x + y_T$

EQUIL LINE:

$y^* = \frac{P}{\pi} = \frac{P}{1} = P = X$

$N_G = 2.84 = \displaystyle\int_{y_T}^{.005} \frac{dy}{y-y^*}$; $y - y^* = x + y_T - x = y_T$

$N_G = 2.84 = \displaystyle\int_{y_T}^{.005} \frac{dy}{y_T} = \frac{.005 - y_T}{y_T}$

$y_T = .001302$

$recovery = \dfrac{.005 - .001302}{.005} \times 100$

$= 73.96\%$

2-2 (con't)

(b)

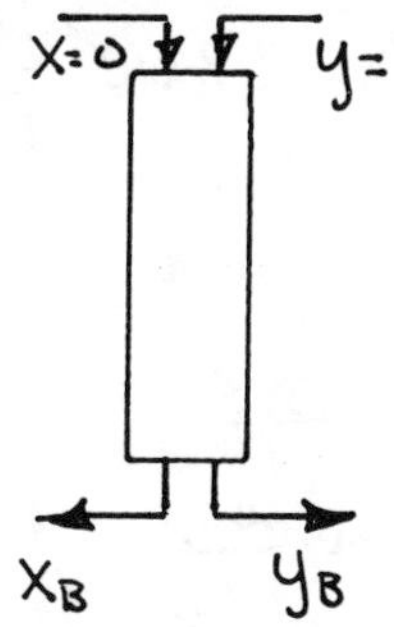

y=.005

cocurrent op line

$y = .005 - x$

EQUIL LINE

$y^* = x$

$N_G = 2.84 = \displaystyle\int_{y_B}^{.005} \frac{dy}{y - y^*}$; $y - y^* = .005 - 2x$

$= 2y - .005$

$N_G = 2.84 = \displaystyle\int_{y_B}^{.005} \frac{dy}{2y - .005}$

$= \frac{1}{2} \ln \dfrac{.005}{2y_B - .005}$

$y_B = .0025$

$recovery = \dfrac{.005 - .0025}{.005} \cdot 100 = 50\%$

2-3

"reasonable" approach: 5°F cocurrent
10°F countercurrent

(a)

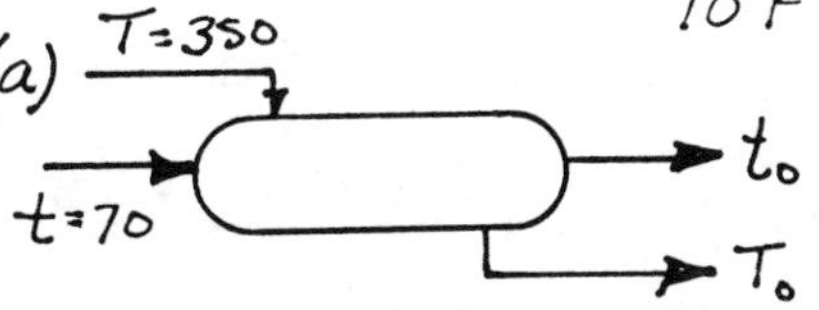

$T_o - t_o = 5$

heat balance

$mc_p\Delta t = mc_p\Delta t$

$10000(1)(350 - T_o) = 10000(1)(t_o - 70)$

$T_o = 5 + t_o$

$(350 - 5 - t_o) = t_o - 70$

$t_o = 207.5°F$

$Q = 10000(1)(207.5 - 70) = 1.375 \times 10^6 \frac{Btu}{hr}$

(b)

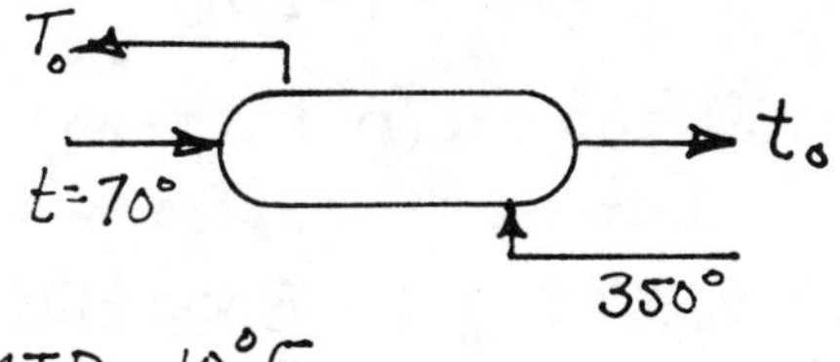

$LMTD = 10°F$

$350 - t_o = T_o - 70 = 10$ $t_o = 340$

2-3 (con't)

$Q = 10000(1)(340-70) = 2.7\times10^6 \frac{Btu}{hr}$

(c)

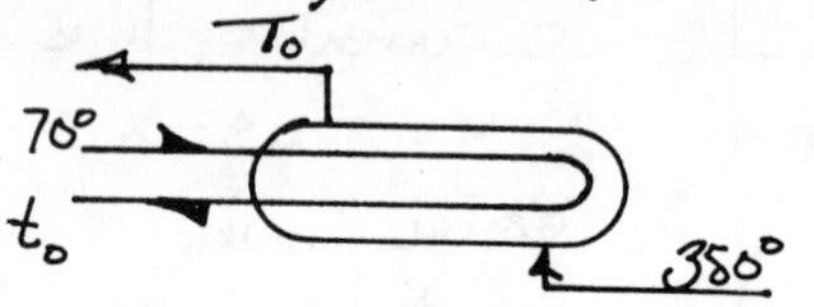

$R = \frac{T_1 - T_0}{t_0 - t_1} = 1$ since $\frac{wC_p}{wC_p} = 1$

$P_{MAX} = .575$; $P = \frac{t_o - t_1}{T_1 - t_1} = \frac{t_o - 70}{350-70} = .575$

$t_o = 231°F$

$Q = 10000(1)(231-70) = 1.61\times10^6 \frac{Btu}{hr}$

2-4

(a) At upset: inlet - outlet = accumulation

$Q \equiv gal/hr.$; $\rho \equiv lb/gal$; $C \equiv \%$

$Q_{in}\rho C_{in} - Q_{out}\rho C_T = \frac{d}{dt}[\rho V C_T]$

$Q_{in}C_{in} - Q_{OUT}C_T = C_T\frac{dV}{dt} + V\frac{dC_T}{dt}$

B.C.: $t=0$, $V_0 = 8000$, $C_T = 0.1$

but $\frac{dV}{dt} = 1000 - 1500 = -500$ gal/hr

$V = V_o - 500t$; $Q_{in} = 1000$ $C_{in} = .05$

$Q_{OUT} = 1500$

$1000(.05) - 1500C_T = C_T(-500) + (8000-500t)\,dC_T/dt$

$50 - 1500C_T = -500C_T + (8000-500t)dC_T/dt$

$\int_0^t \frac{dt}{160-10t} = \int_{.1}^{.098} \frac{dC_T}{1-20C_T}$

$\ln\frac{160-10t}{160} = \frac{1}{2}\ln\frac{1-20(.098)}{1-20(.1)}$

$t = 0.323$ hr (or 19.4 min) before alarm sounds

(b) $V = V_0 - 500t$; $0 = 8000 - 500t$

$t = 16$ hours to empty

2-5

(a) convert to lb dry air/hr:

fresh air (30°F, 10% RH): $v = 12.35\ \frac{ft^3}{lb\,d.a.}$

recycle air (70°F, 50% RH): $v = 13.52\ ft^3/lb\,d.a.$

air to building (72°F, 50% RH): $v = 13.58$

rate: $10^6/13.58 = 73638$ lb d.a./hr.

fresh air

$73638 - 36982 = 36656\ \frac{lb\,d.a.}{hr}$

$= 7545$ cfm

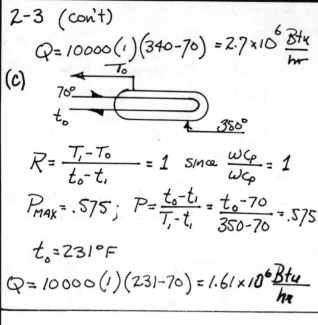

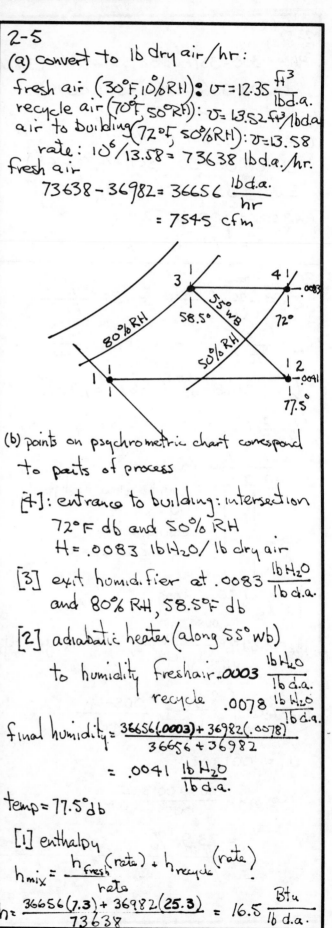

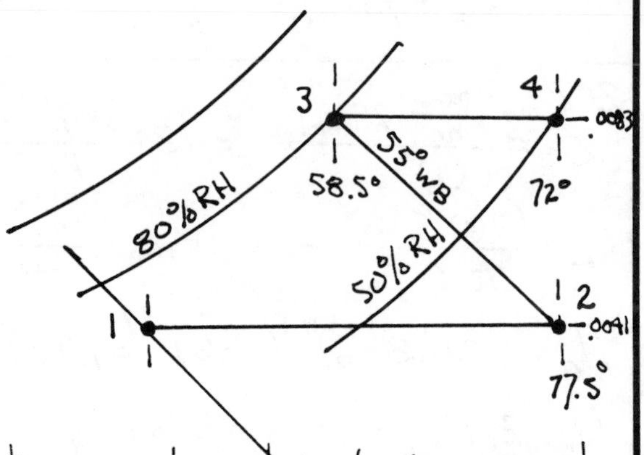

(b) points on psychrometric chart correspond to parts of process

[4]: entrance to building: intersection 72°F db and 50% RH

$H = .0083$ lbH_2O/lb dry air

[3] exit humidifier at $.0083\ \frac{lb\,H_2O}{lb\,d.a.}$ and 80% RH, 58.5°F db

[2] adiabatic heater (along 55° wb) to humidity freshair $.0003\ \frac{lb\,H_2O}{lb\,d.a.}$

recycle $.0078\ \frac{lb\,H_2O}{lb\,d.a.}$

final humidity $= \frac{36656(.0003) + 36982(.0078)}{36656 + 36982}$

$= .0041\ \frac{lb\,H_2O}{lb\,d.a.}$

temp = 77.5° db

[1] enthalpy

$h_{mix} = \frac{h_{fresh}(rate) + h_{recycle}(rate)}{rate}$

$h = \frac{36656(7.3) + 36982(25.3)}{73638} = 16.5\ \frac{Btu}{lb\,d.a.}$

2-5 (cont)

LOCATION	°F	h, lb H_2O/lb d.a.
preheater in	50	.0041
humidify. in	77.5	.0041
humidify. out	58.5	.0083

[c] WATER TEMP = 55°F

[d]

$$\ln \frac{t_{yb} - t_s}{t_{ya} - t_s} = \frac{h_y a V_T}{C_s V} = \frac{90\, V_T}{.244(73638)}$$

$$\ln \frac{77.5 - 55}{58.5 - 55} = \frac{90\, V_T}{.244\,(73638)}$$

$$V_T = 371 \text{ ft}^3$$

2-6 If the operation can be changed so that fewer theoretical plates are required, then the present column will do a better job. The results can be seen on McCabe-Thiele diagrams

[a] [b]

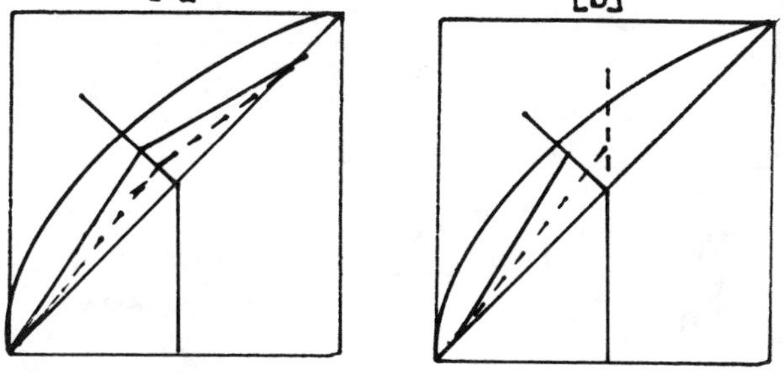

X_F X_F

Present ———

Modified - - - -

Both [a] & [b] can meet the objective provided the feed plate can be relocated.

2-7

(a) $A + B \rightarrow R$

$R + B \rightarrow S$

The concentrations of A, B, R & S are:

$$C_A + C_R + C_S = C_{A0}$$

$$C_B + C_R + 2C_S = C_{B0}$$

WHICH LEADS TO:

$$C_S = [C_{B0} - C_B] - [C_{A0} - C_A]$$

$$C_R = 2[C_{A0} - C_A] - [C_{B0} - C_B]$$

the given data:

t	C_A	C_B
0	1.4	3
10	.89	2.41
15	.74	2.23
20	.64	2.07
25	.55	1.92
30	.47	1.80
40	.37	1.61
50	.29	1.45
60	.23	1.32
70	.19	1.22

$\bar{C}_A$ = **average concentration between** time intervals

$\bar{C}_B$ & $\bar{C}_R$ (same)

calculated:

t	C_S	C_R	$\bar{C}_A$	$\bar{C}_B$	$\bar{C}_R$
10	.08	.43	1.145	2.705	0.215
15	.11	.55	.815	2.32	0.49
20	.17	.59	.69	2.15	0.57
25	.23	.62	.595	1.995	0.605
30	.27	.66	.51	1.86	0.64
40	.36	.67	.42	1.705	0.665
50	.44	.67	.33	1.53	0.67
60	.51	.66	.26	1.385	0.665
70	.57	.64	.21	1.27	0.65

2-7 (con't)

$$-r_A = \frac{\Delta C_A}{\Delta t}$$

$$r_S = \frac{\Delta C_S}{\Delta t}$$

$$K_1 = \frac{r_A}{\bar{C}_A \bar{C}_B}$$

$$K_2 = \frac{r_S}{\bar{C}_R \bar{C}_B}$$

t	$-r_A$	r_S	k_1	k_2
10	.051	.008	.0165	.0138
15	.03	.006	.0159	.0053
20	.02	.012	.0135	.0098
25	.018	.012	.0152	.0099
30	.016	.008	.0169	.0067
40	.01	.009	.014	.0079
50	.008	.008	.0158	.0078
60	.006	.007	.0167	.0076
70	.004	.006	.015	.0073

(b)

$\bar{K}_1$ = .0155 liters/gmole-min

$\bar{K}_2$ = .0085 liters/gmole-min

2-8

$$v = \sqrt{2g_c \Delta H}$$

$\Delta H = 1.4\ (13.6) = 19.04$ in H_2O = 1.59 ft H_2O

$$v = \sqrt{64.4\ (1.59)} = 10.11 \text{ ft/sec}$$

this is centerline velocity

2-8 (con't.)

To find average velocity calc. N_{Re}

$d = 3.068$; $\mu = 1.1$ cp; $\rho = 62.4$; $A = .0513$ ft^2

$$N_{Re} = 123.9\ \frac{3.068(10.11)62.4}{1.1} = 2.2 \times 10^5$$

use Nikuradse's equation

$$v_y = v_{center}(y/r_o)^{1/8}$$

Colebrook showed for turb. flow, v_{avg} occurs at $0.78\,r_o$.

so $v_{avg} = v_{center}(.78r_o/r_o)^{1/8} = .969\,v_{center}$

flow rate $= v_{avg} A \rho = .969(10.11)(.0513)(62.4) = 31.36$ lb/sec

2-9

substituting constants into given rate equation

$$r_A = \frac{0.0696\ (P_A - P_R P_S/0.201)}{(1 + 0.44 P_A + 1.524 P_{R_S})^2}$$

with pure A as feed $P_R = P_S$

$$P_{R_S} = \frac{P_R + P_S}{2} = P_R$$

$$r_A = \frac{0.0696\ (P_A - P_R^2/0.201)}{(1 + 0.44 P_A + 1.524 P_R)^2}$$

design eqtn:

$$\frac{W}{F_{A_0}} = \int_0^{X_A} \frac{dX_A}{r_A}$$

graphical or numerical integration req'd

moles A = $N_{A_0}(1 - X_A)$

moles R = $N_{A_0} X_A$; moles S = $N_{A_0} X_A$

total moles = $N_{a_0} + N_{a_0} X_A = N_{A_0}(1 + X_A)$

$$P_A = \frac{N_{A_0}(1 - X_A)}{N_{A_0}(1 + X_A)}(.1) = 0.1\ \frac{1 - X_A}{1 + X_A}$$

$$P_R = \frac{N_{A_0} X_A}{N_{A_0}(1 + X_A)}(.1) = 0.1\ \frac{X_A}{1 + X_A}$$

2-9 (con't)

X_A	$1/r_A$
0	156.6
.02	163.4
.04	170.6
.06	178.1
.08	186.1
.1	194.5
.12	203.4
.14	212.9
.16	222.9
.18	233.5
.2	244.8

using Simpson's approximation

$I = \int f(x)dx$ for $\Delta x = .02$

$I = \frac{W}{F_{A0}} = 39.31 \frac{\text{lb/cat-hr}}{\text{lbmole}}$

$F_{A0} = 20$

$W = 39.31(20) = 786.3$ lb cat

2-10 100 lb feed : basis

(a) preheater

$$Q = \left[C_{P(liq)}(156-77) + \lambda + C_{P(vap)}(300-156)\right]100$$

where $\lambda = 6900(1.8)\frac{1}{86} = 144.4$

$$Q = \left[.6(79) + 144.4 + .47(144)\right]100 = 25948 \text{ BTU}$$

(b) material balance

	component	lb	lbmole
feed:	n-hexane	100	1.162
products:	n-hexane	50	.581
	cyclohex	48.8	.581
	hydrogen	1.2	.581

Enthalpy balance for reactor

$$\Sigma H_{reactants} + Q_{ADDED} = \Delta H_{Rx} \times \text{moles} + \Sigma H_{PROD}$$

$\Sigma H_{Reactants} = 25950$

$$\Sigma H_{PRODUCTS} = 50(.6(79) + 144.4 + .47(244)) + 48.8(.45(103) + \lambda + .4(220)) + 1.2(3.5)(323) = 30757 \text{ Btu}$$

2-10 (con't)

from heats of combustion

$$\Delta H_{Rx} = (-937800 - 68372) + 994600 = 11572 \text{ cal/gmole} = 20830 \text{ Btu/lbmole}$$

$Q = 20830(.581) + 30757 - 25950$

$Q = 16909$ Btu added

2-11

(a) $h_f = f\left(\frac{L}{D}\right)\frac{v^2}{2g_c}$

$\mu = 1.16$ cp

$v = 4.82$ ft/sec

$$N_{Re} = \frac{(6.065/12)(62.33)(4.82)}{1.16 \times 6.72 \times 10^{-4}} = 1.76 \times 10^6$$

$\varepsilon/D = .001$

$f = .0196$ from chart

$$\frac{L}{D} = \frac{350 + 2050 + 815\left(\frac{6.065}{12}\right)}{6.065/12} = 5563.6$$

$$h_f = .0196(5563.6)(4.82)^2/(2 \times 32.2)$$

$h_f = 39.3$ ft-lb_f/lb_m

Shaft work

$$\frac{v^2}{2g_c} + \frac{(z_1 - z_2)g}{g_c} + h_f + \eta h_{turbine} = 0$$

$$\frac{4.8^2}{2(32.2)} + (3750 - 3870)\frac{32.2}{32.2} + 39.3 = \eta h_{turbine}$$

$+\eta h_{turbine} = +80.3$ ft-lb_f/lb_m

2-11 (con't)

(b)

$$\text{power} = 80.3\ \frac{\text{ft-lb}_f}{\text{lb}_m} \times 58\ \frac{\text{ft}^3}{\text{min}} \times 62.33\ \frac{\text{lb}_m}{\text{ft}^3} \times \frac{1}{60}\ \frac{\text{min}}{\text{sec}} \times 1.356\ \frac{\text{watts-sec}}{\text{ft-lb}_f}\ (.86)$$

= 5642 watts

= 56.4 100-watt bulbs

2-12

Determine U_o clean

$$\frac{1}{U_D} = \frac{1}{U_o} + \text{dirt factor}$$

$$\frac{1}{292} = \frac{1}{U_c} + .0022$$

$U_c \cong 820$ Btu/hr °F ft²

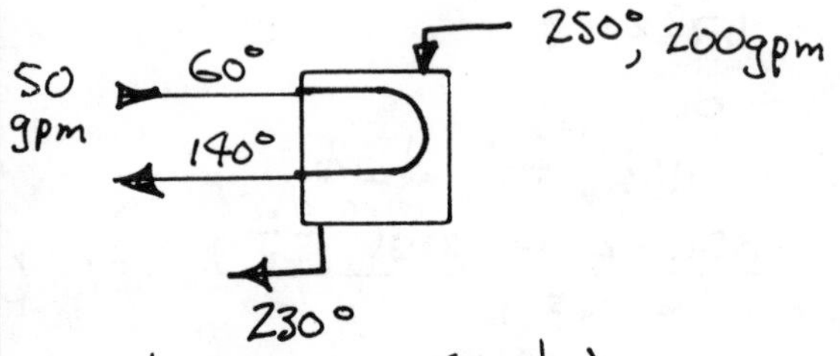

heat balance (dirty)

$$W_1(140-60) = W_2(250-T_2)$$

$$\frac{W_1}{W_2} = \frac{250-T_2}{80} \qquad T_2 = 230°$$

$$\text{LMTD} = \frac{(250-140)-(230-60)}{\ln \frac{250-140}{230-60}} = 137.8$$

$$R = \frac{T_1-T_2}{t_2-t_1} = 0.25$$

$$P = \frac{t_2-t_1}{T_1-t_1} = .421$$

1-2 Xchr $F_T = 0.985$

2-12 (con't)

LMTD = .985 (137.8) = 135.7

$$Q_{DIRTY} = Wc_p\Delta t = (50)(8.33)(62.33)(1.0)(80)$$

2.08×10^6 Btu/hr

$$Q = UA(LMTD) = 292(A)(135.7) = 2.07\times10^6$$

$A = 52.2\ \text{ft}^2$

To find service other than design use Ten Broecke Charts

$$UA/wc = 820(52.2)/[50(8.33)60] = 1.71$$

$R = 0.25$

from chart

$P = 0.73$ for 1-2 exchanger

$$0.73 = \frac{t_2-60}{250-60}; \quad t_2 = 198.7°\text{F}$$

heat Balance

$$T_2 = 250 - \frac{1}{4}(198.7-60) = 215.3°\text{F}$$

2-13

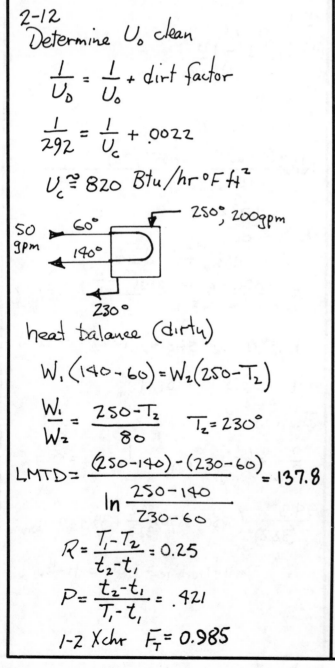

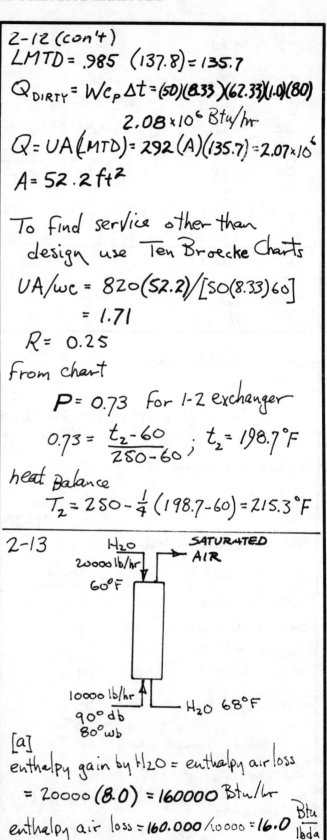

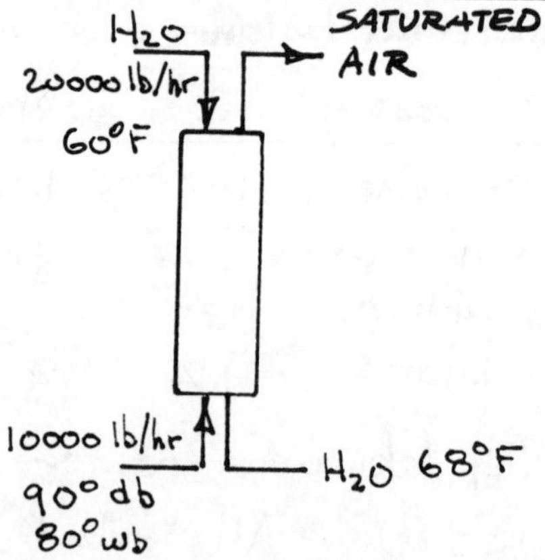

[a]

enthalpy gain by H_2O = enthalpy air loss

= 20000 (8.0) = 160000 Btu/hr

enthalpy air loss = 160.000/10000 = 16.0 $\frac{\text{Btu}}{\text{lbda}}$

Exit air enthalpy = 43.69 − 16.0

= 27.69 Btu/lb d.a.

2-13 (con't)

Saturated, humidity chart:

exit air temp = 61.8 °F

exit:

$h = 0.0118$ lb H_2O/lb d.a.

entering:

$h = 0.0198$ lb H_2O/lb d.a.

[b]

H_2O condensed $= 1000(0.0198 - 0.0118) = 8.0$ lb

2-14

$$\frac{V_I}{V_D} = \text{volumetric efficiency} = 1 + C - C\left(\frac{P_2}{P_1}\right)^{1/k}$$

$$C = \frac{\text{clearance vol.}}{V_D}$$

V_I = intake volume

V_D = displacement volume

$$\frac{V_I}{V_D} = 1 + 0.05 - 0.05\left(\frac{6}{1}\right)^{1/1.4} = .8702$$

$V_I = 100$ cfm; $V_D = 100/.8702 = 114.9$ cfm

$$\text{displacement} = 114.9 \times \frac{1}{200\text{ strokes}} \times \frac{1\text{ stroke}}{2\text{ displacements}}$$

$$= .287\ ft^3$$

Ideal gas

$$\text{theo. work} = -\frac{K}{K-1} RT_1\left[1 - \left(\frac{P_2}{P_1}\right)^{\frac{K-1}{K}}\right]$$

$T = 90°F = 550°R$

$$\text{theo work} = -\frac{1.4}{.4}(1.987)550\left[1 - 6^{.4/1.4}\right]$$

$= 2557$ Btu/lbmole

$$\eta = \frac{PV}{RT} = \frac{(1\text{ atm})(100\ ft^3)}{(.73)(550°)} = .249\ \frac{\text{lbmoles}}{\text{min}}$$

2-14 (con't)

$$\text{theo hp} = 2557\ \frac{\text{Btu}}{\text{lbmole}} \times .249\ \frac{\text{lbmole}}{\text{min}} \times \frac{1}{60}\ \frac{\text{min}}{\text{sec}}$$

$$\times 778\ \frac{\text{ft-lb}}{\text{Btu}} \times 1\ \frac{\text{hp}}{550\text{ ft-lb/sec}}$$

$= 15.01$ hp

actual $= 15.01/0.82 = 18.3$ hp

2-15

(a) For N equal sized CSTR's in series for 1st order reaction

$$\frac{C_0}{C_N} = [1 + K\tau]^N$$

for constant feed rate τ is proportional to reactor volume.

Since $C_0/C_N = \dfrac{1}{1 - .95} = 20$

$V = v_0\tau$ v_0 = flow rate

N	τ	V	FCI
1	19/K	$19v_0/K$	6.779 ϕ
2	3.472/K	$3.472v_0/K$	4.492 ϕ
3	1.714/K	$1.714v_0/K$	4.259 ϕ
4	1.115/K	$1.115v_0/K$	4.293 ϕ

where $FCI = NAV^{.65} = N\phi$ (constant)

$$\phi = A(v_0/K)^{.65}$$

N = 3 is optimum number in series

(b) Parallel arrangement is not optimum because $V_{each} = V_{single}/N$. For instance, for parallel $N = 2$, $V = 9.5\,V_0/K$ and

$$FCI_{2\,PARALLEL} = 8.641\,\phi$$

Use 3 equal sized CSTRs in series.

2-16

[a] since the fiber is insoluble, all fiber in feed will leave in last stage raffinate

$\therefore f_n = 20$ lb/hr

[b] by material balance, if 98% is to be removed, 2% remains

$\therefore \sigma_n = 0.02(80) = 1.6$ lb/hr

[c] on a solvent-free basis there will be $\frac{1.6}{(20+1.6)} = .0741$ fraction oil. actual composition lies on underflow locus. Draw line of constant oil to fiber ratio by drawing from 0.0741 at zero solvent to pure solvent apex. Intersection of this line and underflow locus is desired point

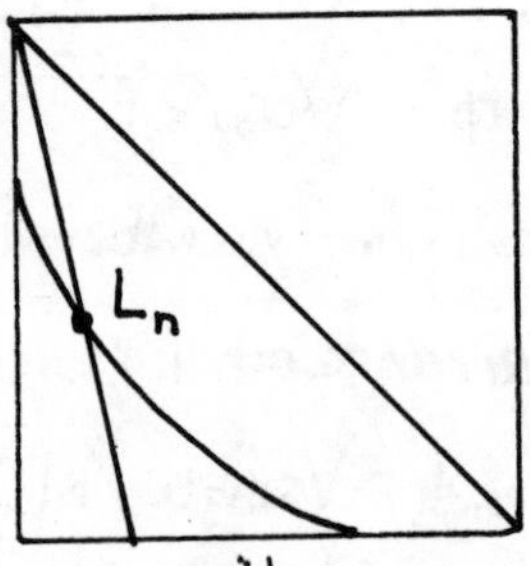

At L_n, oil = 4.1%
solvent = 45.5%
fiber = 50.4%

solvent at $L_n = \frac{45.5}{50.4} \times 20 = 18.1 \frac{\text{lb}}{\text{hr}}$

2-16 (con't)

[d] minimum solvent rate is represented by intersection of line from pure solvent to L_o and line from L_n to pure oil

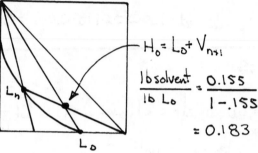

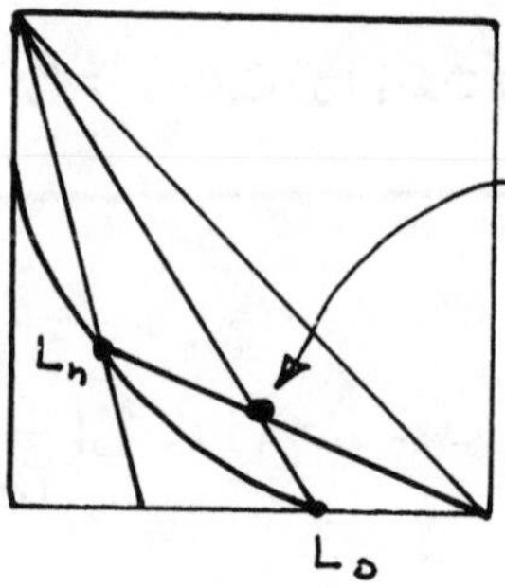

$H_o = L_o + V_{n+1}$

$\frac{\text{lb solvent}}{\text{lb } L_o} = \frac{0.155}{1-.155} = 0.183$

$V_{min} = 0.183 \times 100 = 18.3$ lb/hr

[e] $V = 18.3(1.3) = 23.8$ lb/hr

% solvent = 23.8/123.8 = 19.2%

feed lies along line from L_o at 19.2% to pure solvent (point F). Extension from L_n through F to 45° line gives V_1. From V_1 to L_o and from solvent to L_n intersects at Δ. Draw from V_1 to pure fiber. From Δ through intersection of underflow and $V_1 \rightarrow$ pure fiber. This is V_2

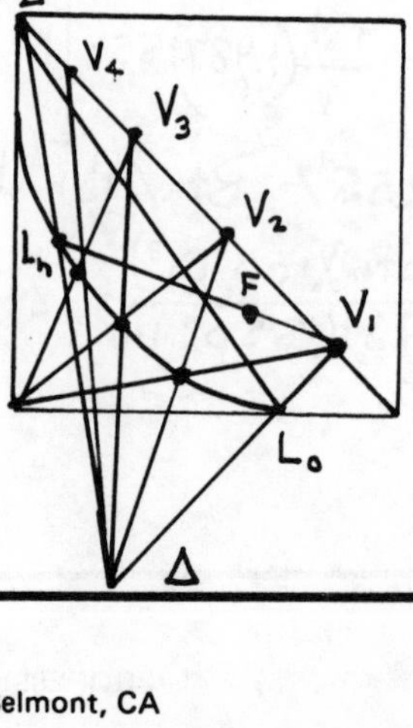

graphical construction yields 8 theo. stages

2-17

$$K t C_{A_0}(M-1) = \ln \frac{M - X_A}{M(1-X_A)} \;;$$

$$M = C_{B_0}/C_{A_0} = \frac{.0852/2}{.0912/2} = .9342$$

$$C_{A_0} = .0912/2 = .0456$$

$$X_A = 1 - C_A/C_{A_0}$$

t	C_A	$X_A\left[=1-\frac{C_A}{.0456}\right]$	K
90	.0373	.1820	.0585
180	.0291	0.3618	.0754
270	.0231	0.4934	.0877
615	.0134	0.7061	.1005
915	.0122	0.7325	.0780
1515	.0081	0.8224	.0868
1815	.0063	0.8618	.1062

$$K = \frac{1}{t C_{A_0}(M-1)} \ln \frac{M - X_A}{M(1-X_A)}$$

$$K = \frac{1}{t(.0456)(.9342-1)} \ln \frac{.9342 - X_A}{.9342(1-X_A)}$$

$$K = \frac{-333.28}{t} \ln \frac{.9342 - X_A}{.9342(1-X_A)}$$

AVERAGE* $K = .0891 \ \frac{\text{liter}}{\text{gmole-sec}}$

* not including 90 sec data point (std dev = .0122)

2-18

$$NPSH = h_s - h_f - p^\circ$$

at 180°F $p^\circ = 7.51$ psi $= 17.3$ ft H_2O

$h_s = 14.7$ psi $+ 5$ ft $= 33.9$ ft $+ 5$ ft $= 38.9$ ft H_2O

$$N_{Re} = 50.6 \frac{(200)(60.6)}{4.026(0.35)} = 4.35 \times 10^5$$

$f = .0152$ (from chart)

$$h_f = 0.0311 \frac{f L_e Q^2}{d^5}$$

$$L_e = 8 + L_{ELBOWS} = 8 + 2\left(40\left(\tfrac{4}{12}\right)\right) = 34.7 \text{ ft}$$

$$h_f = .0311 \frac{(.0152)(34.7)(200)^2}{4.026^5} = .62 \text{ ft}$$

contraction loss:

$$h_{CL} = K\frac{v^2}{2g} = 0.4(1.25)\frac{5.04^2}{64.4} = 0.197 \text{ ft}$$

$$NPSH = 38.9 - 17.3 - 0.62 - 0.197 = 20.8 \text{ ft}$$

2-19

from steam tables at 5" Hg

$t = 133.8$°F $\quad h_{fg} = 1017.7 \ \frac{\text{Btu}}{\text{lb}}$

Freon 12 at 25 psia

$t = 2.2$°F $\quad h_f = 8.99$

$h_g = 77.5$

$$\frac{1}{U_o} = \frac{1}{h_o} + \frac{1}{h_{io}} = \frac{1}{h_o} + \frac{1}{h_i A_i/A_o}$$

for 3/4-inch 18 gage tube OD = .75

$A_i = .1707$ ft²/ft $\quad h_o = 1500$

$A_o = .1963$ ft²/ft $\quad h_i = 500$

2-19 (con't)

$$\frac{1}{U_o} = \frac{1}{1500} + \frac{1}{500(.1707/.1963)}$$

$U_o = 337$ Btu/hr ft^2°F

$Q = UA\Delta t = 337(0.1963)(133.8 - 22.2)$

$Q = 8706$ Btu/hr-ft of tubing

steam at 5" Hg $\lambda = 1017.7$

[a] steam condensed $= \frac{8706}{1017.7} = 8.55 \frac{\text{lb}}{\text{hr-ft}}$

[b] Freon-12 @ 40% quality

$h = 0.40(77.5) + 0.6(8.99) = 36.39 \frac{\text{Btu}}{\text{lb}}$

at 95% quality

$h = 0.95(77.5) + 0.05(8.99) = 74.07 \frac{\text{Btu}}{\text{lb}}$

10 ft section

$Q = 8706(10) = 87060 \frac{\text{Btu}}{\text{hr}}$

Freon-12 flow

$$\frac{87060}{74.07 - 36.39} = 2310 \frac{\text{lb}}{\text{hr}}$$

2-20

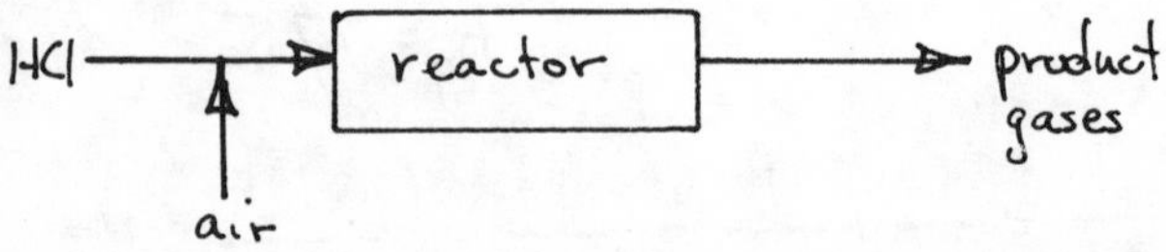

Basis: 100 gmoles product gas

2-20 (con't)

	PRODUCT		FEED	
	gmoles	mole %	gmoles	mole %
HCl	12.04	12.04	40.14	37.50
N_2	52.83	52.83	52.83	49.36
O_2	7.03	7.03	14.06	13.14
Cl_2	14.05	14.05	0	0
H_2O	14.05	14.05	0	0
tot.	100.	100	107.03	100

Enthalpy balance

$\Sigma H_{feed} + Q = \Sigma H_{PROD} + \Delta H^\circ_{25} \times \text{moles}$

$\Sigma H_{feed} = 0$ (25°C basis)

use $\bar{c}_p$ to get products

component	gmoles	$\bar{c}_p$	H cal/100 mole Prod
HCl	12.04	7.02	35900
Cl_2	14.05	8.57	51200
O_2	7.03	7.51	22400
N_2	52.83	7.13	160100
H_2O	14.05	8.4	50200
total			319800

$\Sigma H_{PROD} \cong 319800 \frac{\text{cal}}{100 \text{ moles Prod}}$

$4HCl + O_2 \rightarrow 2Cl_2 + 2H_2O$

$\Delta H^\circ_{25} = 2(-57801) + 2(0) - [4(-22060) - 0]$

$= -27360$ cal/4 gmole HCl

$\Delta H^\circ_{25} = -6840$ cal/gmole HCl

$\therefore 0 + Q = 319800 - 6840(28.1)$

$= \frac{127600 \text{ cal added}}{100 \text{ gmole prod gas}}$

3-1

$$\Delta P = h\rho = [0.6-(-0.5)]\ \frac{1}{12}\frac{\text{ft}}{\text{in}} \times 62.4 \frac{\text{lb}}{\text{ft}^3}$$

$$\Delta P = 5.72\ \text{lb/ft}^2$$

$$\rho_{air} = \frac{1}{359}\frac{\text{lbmole}}{\text{ft}^3} \times 29 \frac{\text{lb}}{\text{lbmole}} \times \frac{492°\text{R}}{760°\text{R}} \times \frac{730\ \text{mmHg}}{760\ \text{mmHg}}$$

$$\rho_{air} = 0.05023\ \text{lb/ft}^3$$

$$\frac{v_{max}^2}{2g_c} = \frac{\Delta P}{\rho} = h_v$$

$$v = \sqrt{2 g_c \Delta P/\rho}$$

$$v_{max} = \sqrt{2(32.2)(5.72)/.05023}$$

$$v_{max} = 85.6\ \text{ft/sec}$$

$$v_{avg}/v_{max} = 0.81$$

$$v_{avg} = 0.81(85.6) = 69.3\ \text{ft/sec}$$

$$Q = vA = 69.3\ \frac{\pi}{4} 4^2 = 870.8\ \text{ft}^3/\text{sec}$$

$$SO_2 \text{ rate} = 870.8(.05023)\left(3600\frac{\text{sec}}{\text{hr}}\right)\left(\frac{1}{29}\frac{\text{lbmole}}{\text{lb}}\right)\left(64\frac{\text{lb }SO_2}{\text{lbmole}}\right)(.005)$$

$$= 1737.5\ \text{lb } SO_2/\text{hr}$$

3-2

$$P_A = \gamma_A P_A^\circ X_A;\quad P_W = \gamma_W P_W^\circ X_W$$

$$760 = P_A + P_W = \gamma_A P_A^\circ X_A + \gamma_W P_W^\circ X_W$$

at bubble point X_A & X_W do not change

$X_W = 0.3 \quad X_A = 0.7$

$$\log_{10}\gamma_A = 0.24(.3)^2;\quad \gamma_A = 1.051$$

$$\log_{10}\gamma_W = 0.35(.7)^2;\quad \gamma_W = 1.484$$

3-2 (con't)

at bubble point:

$$760 = 1.051 P_A^\circ(.7) + 1.484 P_W^\circ(.3)$$

$$760 = 0.7357 P_A^\circ + 0.4452 P_W^\circ$$

t	P_A°	P_W°	$P_A + P_W$
100	429.2	760.0	654.1
110	595.1	1074.7	916.3
104	490.4	875.2	750.4
104.4	496.9	887.5	760.7

[a] bubble point 104.4°C

[b] @ 105°C $P_A^\circ = 506.7$

$P_W^\circ = 906.1$

for flash vap.

$$y_i = \frac{z_i\ (\gamma P_i^\circ/P)}{V\left(\frac{\gamma_i P_i^\circ}{P}\right) + 1 - V}$$

$$x_i = (z_i - V y_i)/L$$

$$y_A = \frac{0.4667\ \gamma_A}{0.667\ \gamma_A V + 1 - V}$$

$$y_W = \frac{0.3577\ \gamma_W}{1.1923\ V \gamma_W + 1 - V}$$

$V = 100(.13) = 13$ lbmoles/hr } from table below

$L = 100 - 13 = 87$ lbmoles/hr }

V	X_A	γ_A	γ_W	y_A	y_W	Σy	
.1	.7	1.051	1.484	.5056	.493	.9985	recalc V
.12	.7265	1.042	1.530	.505	.498	1.003	recalc γ
.12	.7265	1.042	1.529	.505	.498	1.003	recalc γ
.13	.726	1.042	1.529	.506	.494	1.000	recalc V

$X_A = .728 \quad X_W = .272$

$y_A = .506 \quad y_W = .494$

3-3

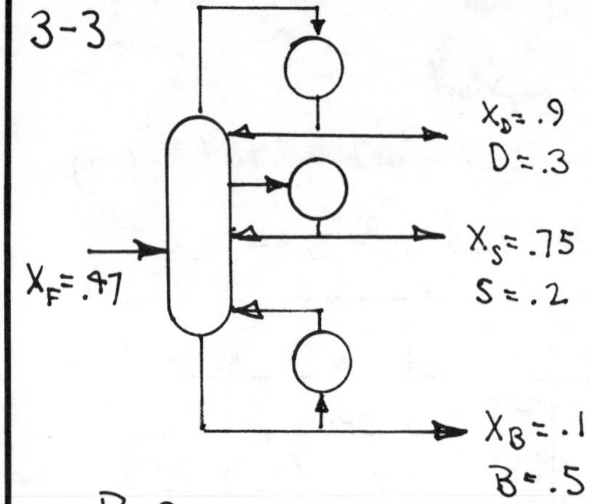

$R = 2$

Reetification:

$$y = \frac{R}{R+1}x + \frac{X_D}{R+1}$$

$$y = \frac{2}{3}x + \frac{.9}{3}$$

$$y = 0.667x + 0.3$$

$$V = (R+1)D = 3(.3) = 0.9$$

$$L = R(D) = 2(.3) = 0.6$$

Intermediate:

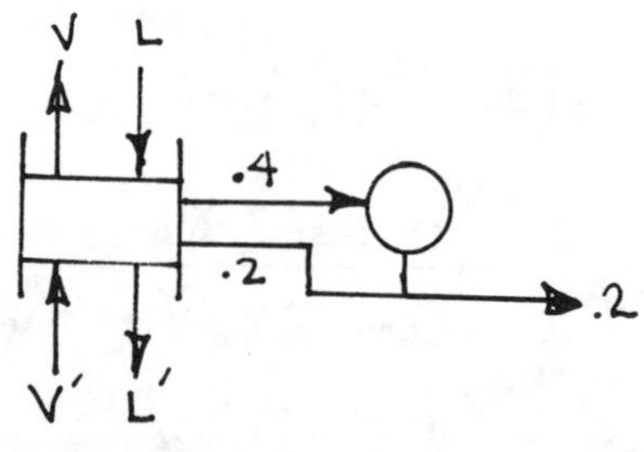

$$V' = V + .4 = 0.9 + .4 = 1.3$$

$$L' = L + .2 = 0.6 + .2 = 0.8$$

$$V'y = L'x + SX_s + DX_D$$

$$y = \frac{L'}{V'}x + \frac{SX_S + DX_D}{V'} = \frac{.8}{1.3}x + \frac{.2(.75) + .3(.9)}{1.3}$$

$$y = 0.615x + 0.32$$

Strip:

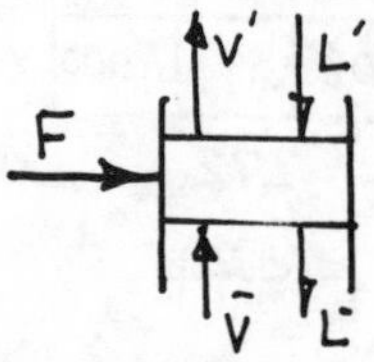

3-3 (con't)

$$V' = \bar{V} = 1.3$$

$$\bar{L} = L' + F = .8 + 1 = 1.8$$

$$y = \frac{\bar{L}}{\bar{V}}x - \frac{BX_B}{\bar{V}} = \frac{1.8}{1.3} - \frac{0.5(0.1)}{1.3}$$

$$y = 1.38x - .0385$$

From plot:

The rectification op line and intermediate op line intersect below the q-line and strip section op line. There are no stages for intermediate section. The column cannot operate at these conditions if the feed plate is below the intermediate section.

3-4

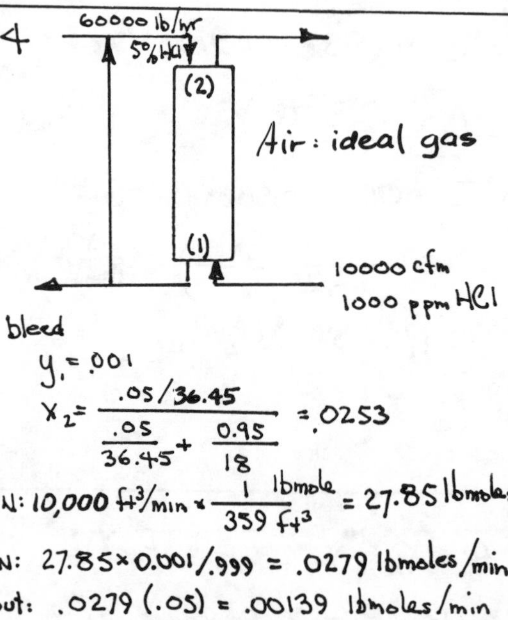

$$y_1 = .001$$

$$x_2 = \frac{.05/36.45}{\frac{.05}{36.45} + \frac{0.95}{18}} = .0253$$

AIR IN: $10{,}000\ ft^3/min \times \frac{1\ lbmole}{359\ ft^3} = 27.85\ lbmoles/min$

HCl IN: $27.85 \times 0.001/.999 = .0279\ lbmoles/min$

HCl out: $.0279(.05) = .00139\ lbmoles/min$

$$y_2 = \frac{.00139}{.00139 + 27.85} = 5 \times 10^{-5}$$

HCl in inlet H_2O = $60000 \frac{lb}{hr} \times \frac{1}{60} \frac{hr}{min} \times .05 \frac{lb\,HCl}{lb\,inlet} \times \frac{1}{36.45} \frac{lbmole}{lb}$

$$= 1.372\ lbmoles/min$$

water in inlet = $60000 \times \frac{1}{60} \times .95/18 = 52.8\ \frac{lbmole}{min}$

HCl in outlet liq = $1.372 + (.0279 - .00139) = 1.4\ \frac{lbmoles}{min}$

$$x_1 = \frac{1.4}{1.4 + 52.8} = .0258$$

PROBLEM 3—3

McCabe—Thiele Plot

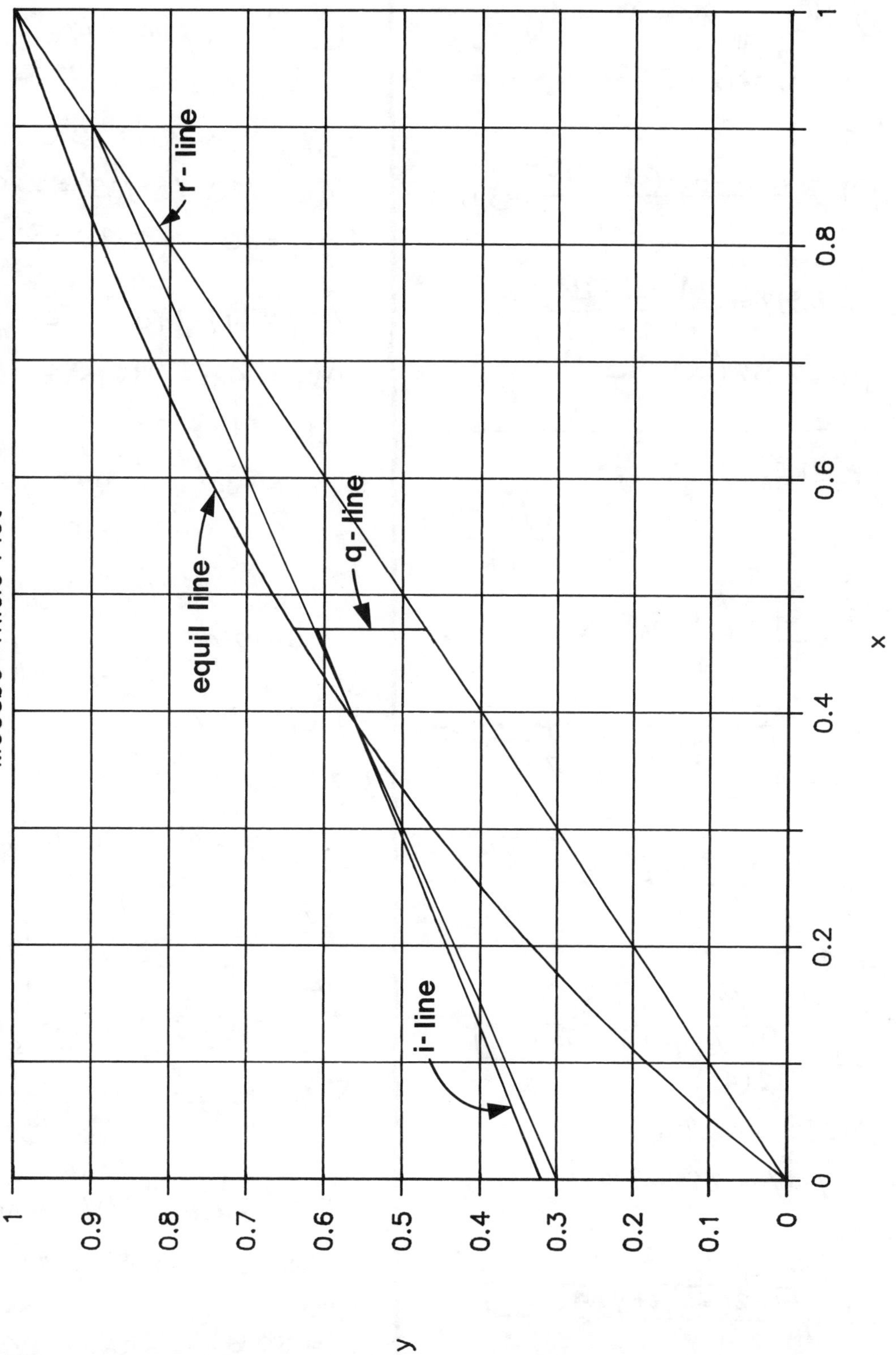

3-4 (con't)

at 5% HCl $P_{HCl} = 4\times10^{-3}$ mmHg

$y_1^* = 4\times10^{-3}/760 = 5.26\times10^{-6} \approx 0$

$y_2^* \cong 0$

$$(y_1 - y^*)_{lm} = \frac{(.001-0)-(5\times10^{-5}-0)}{\ln \frac{10^{-3}}{5\times10^{-5}}} = 3.17\times10^{-4}$$

$$N_{OG} = 2.997 = \frac{yA,out - yA,in}{(y^A - y)\,lm}$$

height $= 2.997(1.5) = 4.5$ ft

3-5 $-r_A = kC_A^2$ $\quad \tau_1 = \frac{V_1}{v_o}$

1st Rxr:

$$k\tau_1 = \frac{C_{A_0} - C_{A_1}}{C_A^2} \qquad C_{A_1} = \frac{-1+\sqrt{1+4\tau_1 k C_{A_0}}}{2\tau_1 k}$$

2" Rxr:

$$C_{A_2} = \frac{-1+\sqrt{1+4\tau_2 k C'_{A_0}}}{2\tau_2 k}$$

since $C'_{A_0} = C_{A_1}$

$$C_{A_2} = \frac{-1+\sqrt{1+4\tau_2 k\left[\frac{-1+\sqrt{1+4\tau_1 k C_{A_0}}}{2k\tau_1}\right]}}{2\tau_2 k}$$

$$C_{A_2} = \frac{-1+\sqrt{1+2\tau_2/\tau_1\left[-1+\sqrt{1+4\tau_1 k C_{A_0}}\right]}}{2\tau_2 k}$$

for 3→12

$\tau_1 k = 10;\ \tau_2 k = 40 \quad \tau_2/\tau_1 = 4$

$$C_{A_2} = \frac{-1+\sqrt{1+2(4)\left[-1+\sqrt{1+4(10)}\right]}}{2(40)}$$

$C_{A_2} = .070627$ moles/liter

$X_{overall} = 1 - .070627/1 = 0.92937$

for 12→3

$\tau_1 k = 40 \quad \tau_2 k = 10 \quad \tau_2/\tau_1 = 0.25$

$$C_{A_2} = \frac{-1+\sqrt{1+2(.25)\left[-1+\sqrt{1+4(40)}\right]}}{2(10)}$$

$C_{A_2} = .0808080$

$X_A = 1 - .0808080/1 = 0.91919$ moles/liter

Smaller reactor first yields higher conversion

3-6

t_s at 115 psia = 170°C

t_s at 135 psia = 177°C

$\Delta t_1 = 170 - 140 = 30°C$
$\Delta t_2 = 140 - (5+105) = 30°C$
$\Delta t_3 = 105 - (5+70) = 30°C$
} initial

$\Delta t_1 = 177 - 140 = 37°C$

$\Delta t_2 = 140 - (119+5) = 16°C$

$\Delta t_3 = 119 - (77+5) = 37°C$

1. Not fouling because
   - steam consumption would not increase
   - second effect temp. diff. dropped.
2. must be steam leak in effect 2.

3-7 Bernoulli:

$$\Delta z + \frac{\Delta p}{\rho} + \frac{\Delta v^2}{2g} + h_f + h_{pump} = 0$$

($\Delta p/\rho$ → 0, $\Delta v^2/2g$ negl, h_{pump} → 0)

$\Delta z = h_f$ {gravity flow}

$$\Delta z = 5 + z = f\sum\left(\frac{L}{D}\right)\frac{v^2}{2g}$$

$$h_f = .00259 f \sum \frac{L}{D} Q^2/d^4 \qquad Q \equiv \text{gpm},\ d \equiv \text{inches}$$

$$\sum\frac{L}{D} = \frac{z}{1.049/12} + \left(\frac{L}{D}\right)_{pipe} + \left(\frac{L}{D}\right)_{XCHR} + \left(\frac{L}{D}\right)_{VALVE} + \left(\frac{L}{D}\right)_{ELBOW} + \left(\frac{L}{D}\right)_{\frac{1}{2}\text{OPEN GATE}}$$

$$= 11.44z + 1559.5 = 11.44z + \frac{90}{1.049/12} + 30 + 340 + 160$$

$$N_{Re} = 50.6\frac{Q\rho}{d\mu} = 50.6(20)(62.4)/(1.049(1))$$

$N_{Re} = 60199 \quad f = .0255$

$$5 + z = .00259(.0255)(11.44z + 1559.5)20^2/1.049^4 = 34.02 + 0.2496z$$

$z = 38.7$ ft

3-8

Fugacity of each component must be equal in all phases

vapor: $f_i = y_i \phi_i P$

where: $\ln \phi_i = \left[2 \sum_{j=1}^{m} (y_j B_{ij}) - B_{mix}\right] \frac{P}{RT}$

liquid:

$$f_i = \gamma_i x_i \phi_{is} P_i^\circ \exp\left[\frac{V_A^2}{RT}(P - P_i^\circ)\right]$$

where: $B_{mix} = y_1^2 B_{11} + 2y_1 y_2 B_{12} + y_2^2 B_{22}$

$$B_{mix} = (.9009)^2(-1304) + 2(.9009)(.0991)(-1169) + (.0991)^2(-1043) = -1277 \text{ ml/gmole}$$

$$\ln \phi_1 = \left[2[(.9009)(-1304) + .0991(-1169)] - (-1277)\right] \frac{.9776}{(82.057)(341.1)}$$

$\phi_1 = .9555$

$\phi_2 = .96446$

$f_{1\,vapor} = (.9009)(.9555)(.9776) = .8415$

$f_{2\,vapor} = (.0991)(.9645)(.9776) = .09344$

for component 1:

$$\exp\left[\frac{V_A^2}{RT}(P - P_i^\circ)\right] = \exp\left[\frac{1}{.617} \times 86.17 \times \frac{1}{82.057} \times \frac{1}{341.1}(.9776 - .9808)\right] = .99998$$

for component 2

$$\exp\left[\frac{V_A^2 (P - P_2^\circ)}{RT}\right] = 1.0010$$

$f_{1\,liq} = 1.0033(.9010)(.9555)(.9808)(.99998) = .8470$

$f_{2\,liq} = 1.3964(.0991)(.9750)(.6793)(1.0010) = .09165$

f_1 vapor versus liq: diff = 0.7%

f_2 vapor versus liq: diff = 1.9%

data good.

3-9 Gas ideal, 1/bmole basis

for vent gas $y = 0.8341$ (From psychrometric chart)

water balance

In [reactor + polluted air] = out [scrubber + treatment]

$$\bar{v}_{vent} = 359 \frac{460 + 203}{492} = 483.8 \text{ ft}^3/\text{lbmole}$$

$$\text{vent total} = 500 \frac{\text{ft}^3}{\text{min}} \times \frac{1}{483.8} \frac{\text{lbmole}}{\text{ft}^3} = 1.034 \frac{\text{lbmole}}{\text{min}}$$

$$\text{vent water} = 1.034(.8341)(18) = 15.52 \frac{\text{lb}}{\text{min}}$$

$$\text{vent air} = 1.034(.1659)(29) = 4.97 \frac{\text{lb}}{\text{min}}$$

polluted air @ 80°F 50% RH

$\rho = 13.8$ ft³/lb dry air
H = .011 lb H_2O/lb d.a. } chart

$$1200/13.8 = 86.96 \frac{\text{lb d.a.}}{\text{min}} + .96 \frac{\text{lb } H_2O}{\text{min}}$$

$$\text{Water in} = 15.52 + .96 = 16.48 \frac{\text{lb}}{\text{min}}$$

Air in = 86.96 + 4.97 = 91.93 lb/min

Out scrubber 72°F @ sat

0.017 lb H_2O/lb d.a.

Air: 91.93 lb/min

H_2O: 1.56 lb/min

to treatment = 16.48 − 1.56 = 14.92 lb/min

3-10 (a)

ERGUN EQUATION:

FLOW THROUGH BEDS (PACKED)

$$-\Delta P = \frac{f_p \rho L v_0^2 (1-\varepsilon)}{D_p g_c \quad \varepsilon^3}$$

$$f_p = \frac{150(1-\varepsilon)}{N_{Re}^*} + 1.75$$

$$N_{Re}^* = \frac{D_p G}{\mu}$$

if $\rho, \varepsilon, f, \mu$ are **constant**

$$\frac{\Delta P_1}{\Delta P_2} = \frac{(L v_0^2)_1}{(L v_0^2)_2} = \frac{(L G^2)_1}{(L G^2)_2}$$

for constant space velocity

$$\frac{L_1}{L_2} = \frac{G_1}{G_2}$$

$$\therefore \frac{\Delta P_1}{\Delta P_2} = \frac{G_1^3}{G_2^3}$$

but $\frac{\Delta P_1}{\Delta P_2} = 2 = \frac{G_1^3}{G_2^3} = \frac{1330^3}{G_2^3}$

$$G_2 = 1056 \ \frac{\text{lbm}}{\text{hr-ft}^2}$$

check for constant f_p

experimental scale:

$$N_{Re}^* = \frac{.01148(1330)}{0.06290} = 243$$

$$f_p = \frac{150(.6)}{243} + 1.75 = 2.12$$

commercial scale:

$$N_{Re}^* = \frac{.01148(1056)}{0.06290} = 193$$

$$f_p = \frac{150(.6)}{193} + 1.75 = 2.22$$

recalculate G_2

$$\frac{\Delta P_1}{\Delta P_2} = 2 = \frac{2.12 \ 1330^3}{2.22 \ G_2^3}$$

$$G_2 = 1040 \ \text{lbm/ft}^2\text{-hr}$$

3-10 (con't)

(b) $S_V = \frac{G}{L} = \frac{1330}{28.2/12} = 566 \ \frac{\text{lb}_m \text{ feed}}{\text{hr-ft}^3 \text{cat}}$

mol. wgt feed $= .913(28.9) + .087(32)$

$= 29.17$

total feed: $\frac{566}{29.17} = 19.4 \ \frac{\text{lbmoles feed}}{\text{hr-ft}^3 \text{cat}}$

moles CH_3OH: $19.4(.087) = 1.69$ lbmoles

moles formaldehyde: $1.69(.914) = 1.54$ lbmoles

lb formaldehyde: $1.54(30) = 46.3 \ \text{lb}_m$

feed rate $= \frac{3400}{46.3}(566) = 41564 \ \frac{\text{lb}_m}{\text{hr}}$

catalyst: $\frac{41564}{566} = 73.4 \ \text{ft}^3$

depth $= \frac{\text{volume}}{\text{area}}$

depth $= \frac{73.4}{41564/1040} = 1.84$ ft

tubes $= \frac{41564}{1040} \frac{(4)}{\pi(.81/12)^2}$

$= 11{,}168$ tubes

3-11

$$C_{12}H_{26} + 18.5O_2 \rightarrow 12CO_2 + 13H_2O$$

MW: $C_{12}H_{26}$: 170 CO_2: 44 H_2O: 18

O_2: 32 N_2: 28

1000 lb fuel/hr $= \frac{1000}{170} = 5.88 \ \frac{\text{lbmoles}}{\text{hr}}$

20% excess O_2: $O_2 = 5.88(1.2)(18.5)$

$= 130.5 \ \frac{\text{lb moles}}{\text{hr}} \ O_2$

$= 4176 \ \frac{\text{lb}}{\text{hr}} \ O_2$

3-11 (con't)

$N_2 = \frac{79}{21}(130.5) = 490.9 \frac{lbmoles}{hr} N_2$

$= 13745 \frac{lb}{hr} N_2$

Weight feed + air = flue gas

$= 1000 + 4176 + 13745$

$= 18921 \, lb/hr$

heat in superheater transferred

$= 18921(.25)(1550-1150) = 1.8921 \times 10^6 \frac{Btu}{hr}$

at 400 psia: sat steam

$h_f = 424.0 \qquad h_g = 1204.5$

at 95% quality

$h = 0.05(424) + 0.95(1204.5)$

$= 1165.5 \, Btu/lb$

at 10000 lb/hr

$q = 1.8921 \times 10^6 / 10^4 = 189.21 \frac{Btu}{lb}$

$h_{exit} = 1165.5 + 189.21 = 1354.7 \, Btu/lb$

since $P=400$ superheat tables for
$h_{exit} = temp \ 686°F \ (h=1354.7)$

3-12

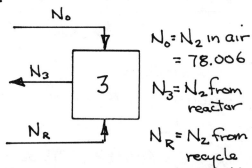

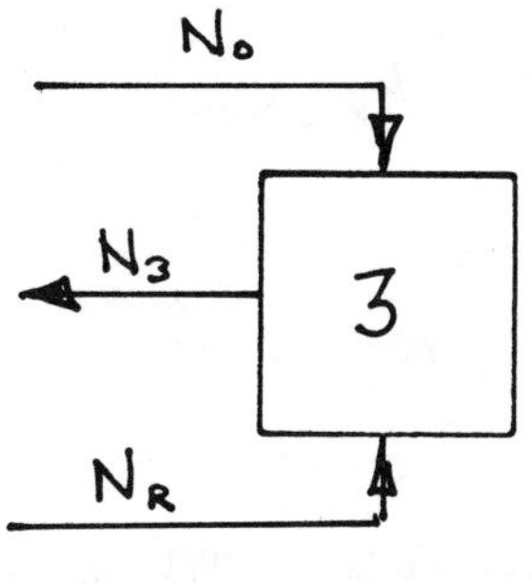

$N_0 = N_2$ in air
$= 78.006$

$N_3 = N_2$ from reactor

$N_R = N_2$ from recycle

$N_w = N_2$ in bleed

Assume
100 moles air

$N_3 = \frac{3}{4}(N_0 + N_R)$
$N_R = 0.9515 \, N_3$ $\Big\}$ $N_0 + N_R = \frac{4}{3} \frac{N_R}{(.9515)}$

3-12 (con't)

$N_R = 194.39$ moles

$H_R = 3N_R = 583.16$ moles H_2

Argon: bleed = feed

$A_w = 0.994$ moles Ar

$N_w = N_R\left[\frac{1}{.9515} - 1\right] = N_R(.05097)$

$N_w = 9.91$ moles N_2

$H_w = (583.16)(0.05097) = 29.72$ moles H_2

$A_w = 0.994$

$(X_A)_R = (X_A)_w = \frac{.994}{9.91 + 29.72 + .994} = .0245$

$A_R = \frac{.994}{1 - .9515} = 20.495$ moles Ar

Recycle:

component	moles	y
N_2	194.39	0.244
H_2	583.16	0.731
Ar	20.5	0.026

3-13

$(rate)^{0.8} \propto$ film thickness

$rate = N_w A = \frac{D_{AB} \, p_A}{RT \Delta z \, P_{lm}} (P_1 - P_2)$

AIR-WATER
$(P_{AIR})_{lm} = \frac{(518-76)-(518-138)}{\ln \frac{442}{380}} = 410 \, mm \, Hg$

$D_{AB} = 0.258(760/518) = 0.379 \frac{cm^2}{sec}$

$\Delta z = \frac{D_{AB} \, p_A}{N_w ART(P_{AIR})_{lm}} (P_{A_1} - P_{A_2})$

$\Delta z = \left(0.379 \frac{cm^2}{sec}\right)\left(518 \, mmHg\right)\left(\frac{1 \, gmole \, °K}{82.06 \, cm^3 \, atm}\right)\left(\frac{1}{298.9°K}\right)\left(\frac{1}{410 \, mm}\right)\left(\frac{1 \, atm}{760 \, mm}\right)$
$\times \left(138-76 \, mmHg\right)\left(\frac{1}{13.19}\right)\left(18 \frac{g}{mole}\right)\left(60 \frac{sec}{min}\right)\left(\frac{1 \, min}{1 \, cm^2}\right)$

$\Delta z = 1.31 \times 10^{-4} \, cm$

3-13 (con't)

n-butylalcohol-air

$$(P_{AIR})_{lm} = \frac{(820-30.5)-(820-54.5)}{\ln\frac{789.5}{765.6}} = 0.777 \text{ mmHg}$$

$$D_{AB} = 0.087(760/820) = 0.0806 \text{ cm}^2/\text{sec}$$

$$\Delta Z_1 = \Delta Z_2 \left(\frac{RATE_1}{RATE_2}\right)^{0.8} = 1.31\times10^{-4}\left(\frac{120}{100}\right)^{0.8}$$

$$\Delta Z_1 = 1.52\times10^{-4}$$

$$\text{EVAP RATE} = \frac{(.0806)(820)(1)}{(82.06)(298.9)(760)(1.52\times10^{-4})}(54.4-30.5)\frac{74.12}{777} \times 60$$

EVAP RATE = 3.19 g/min

3-14

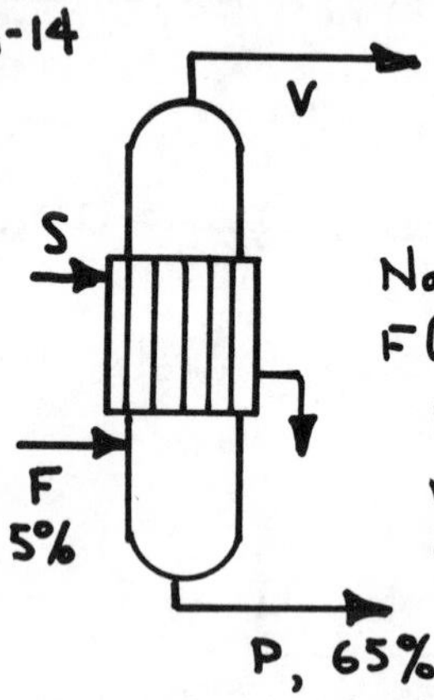

F = V + P

NaOH balance:

F(.05) = V(0.0) + P(0.65)

P = 769.2 lb/hr

V = F − P = 10000 − 769.2

V = 9230.8 lb/hr

assume:
1. C_p NaOH = 1.0
2. neglect B.P.E.
3. adequate U.

from steam tables:

1 psia: $t_{sat} = 101.7°F$; $h_{fg} = 1036 \frac{Btu}{lb}$

40 psig (55 psia): $t_{sat} = 287°F$; $h_{fg} = 919.9 \frac{Btu}{lb}$

for evaporation, heat req'd:

Q = sensible $(t_{feed} \rightarrow t_{evap})$ + latent evap

$Q = mC_p(t_{evap} - t_{feed}) + V\lambda_{evap}$

$Q = 10000(1)(101.7-70) + 9230.8(1036)$

Q = 9880076 Btu/hr

steam:

$Q = UA\Delta t$

since we do not know U, assume worst case, all heat from steam vapor req'd.

$S = Q/(h_{fg} * \text{quality})$

$S = 9880076/(919.9 * .7)$

S = 15343.4 lb/hr steam

now calculate U min

$U = Q/A\Delta t = 9880076/[2500(287-101.7)]$

$U = 21.3$ Btu/ft² °F

This is low and is the minimum value of U for this process. It is likely that U is in the range 100–300 and thus less steam is required, if U is > 21.3.

3-15

(a)

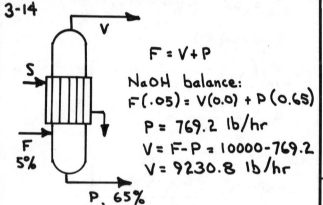

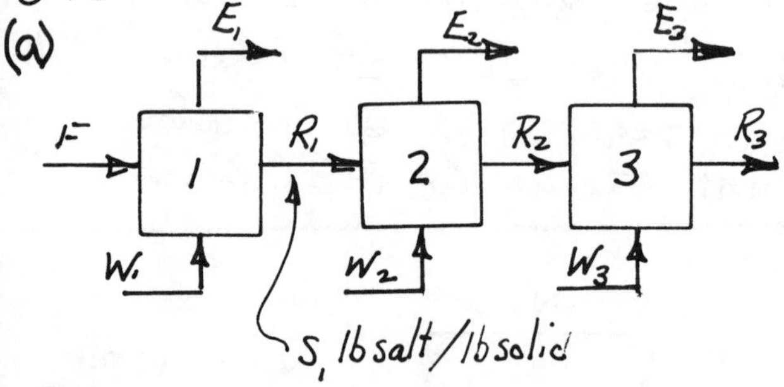

BASIS: F = 1000 lb

F: 300 lb H_2O; 400 lb pulp; 300 lb salt

$W_1 = W_2 = W_3 = 500$ lb

$R_1 = R_2 = R_3 = 600$ lb = 400 lb pulp + 200 lb solution

$E_1 = 900$ lb; $E_2 = E_3 = 500$

$$S_1 = \frac{300}{300+800} = 0.2727 \frac{\text{lb salt}}{\text{lb solution}}$$

R_1: (salt) = 200 lb solution × .2727 = 54.54 lb

3-15(con't)

R_1: (water) = 200 − 54.54 = 145.46 lb

E_1: salt = 300 − 54.54 = 245.46 lb

S_2 = 54.54/(54.54 + 500 + 145.46) = 0.0779

	WATER	PULP	SALT	TOTAL
F	300	400	300	1000
W_1	500	–	–	500
E_1	654.54	–	245.46	900
R_1	145.46	400	54.54	600
S_1				0.2727
W_2	500	–	–	500
E_2	461.04	–	38.96	500
R_2	184.42	400	15.58	600
S_2				0.0779
W_3	500	–	–	500
E_3	488.87	–	11.13	500
R_3	195.55	400	4.45	600
S_3				0.02226

(b) dried solids:

% salt = 100(4.45/404.45) = 1.1 %

% pulp = 100 − 1.1 = 98.9 %

(c) lb salt extracted = (300 − 4.45)/1000 = 0.2956 lb/lb feed = 295.6 lb salt/1000 lb feed

3-16

Let feed = 1 mole/hr

$$y_i = \frac{z_i k_i}{V k_i + (1-V)} \; ; \; x_i = \frac{z_i - V y_i}{L}$$

V = 0.5189 L = 0.4811

i	component	z_i	k_i	y_i	x_i
1	C_2H_6	.002	16.2	.0036	.0003
2	C_3H_8	.040	6.3	.0672	.0107
3	C_4H_{10}	.332	1.35	.3793	.2810
4	C_5H_{12}	.332	0.90	.3152	.3502
5	iC_6H_{14}	.152	0.92	.1459	.1586
6	nC_6H_{14}	.131	0.49	.0873	.1781
7	nC_8H_{18}	.011	.09	.0019	.0208

Σ = 1.0 1.004 0.9997

3-17 for $\varepsilon = 0$ CSTR:

method: integral (guess order, compute k)

guess 1st order

$$\tau = \frac{V}{v_0} \; ; \; kC_A = -r_A$$

$$\tau = \frac{C_{A_0} - C_A}{(-r_A)}$$

v_0	C_A	$C_{A_0} - C_A$	τ	$k\tau$	$k = \frac{C_{A_0} - C_A}{\tau C_A}$
.2	.03	.47	5	15.67	3.13
.4	.057	.443	2.5	7.77	3.11
.8	.102	.398	1.25	3.90	3.12
1.6	.170	.330	0.625	1.94	3.10
3.2	.253	.247	0.3125	0.976	3.12
6.4	.336	.164	0.1563	0.488	3.12

$\bar{k} = 3.117$

$-r_A = 3.117\, C_A$

for PFR, $\varepsilon = 0$

$k\tau = -\ln(1 - X_A)$; $\tau = \frac{1}{3}$; $k\tau = 1.039$

$\ln(1 - X_A) = -1.039$

$X_A = .6462$

3-18

Cooling tower calculations make use of the Merkel enthalpy potential difference relationship

The enthalpy-temperature diagram for the cooling tower on the "hot" day:

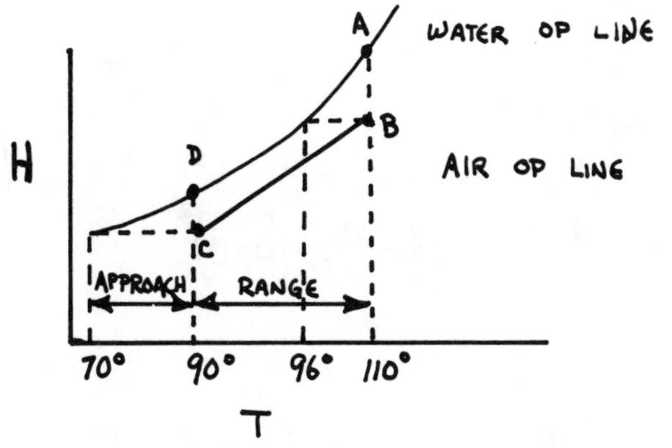

3-18 (con't)

The water op line AD is fixed by saturation enthalpy. The air op line is determined by wet bulb temperature entering and exiting tower.

The Merkel equation

$$\frac{KaV}{L} = \int_{T_1}^{T_2} \frac{dT}{H' - H}$$

The left side of the above equation is the "tower characteristic" and remains unchanged if L/V is unchanged.

Since the water op line has the relationship

$$H = a + bT + cT^2$$

and the air op line is straight so that

$$H = A + BT$$

the tower characteristic can be computed analytically.

$$H' - H = a + bT + cT^2 - (A + BT)$$

$$H' - H = a - A + (b - B)T + cT^2$$

$$\int_{T_1}^{T_2} \frac{dT}{a - A + (b - B)T + cT^2} =$$

$$\frac{2}{\sqrt{q}}\left[\tan^{-1}\frac{2cT_2 + b - B}{\sqrt{q}} - \tan^{-1}\frac{2cT_1 + b - B}{\sqrt{q}}\right]$$

where $q = 4(a - A)c - (b - B)^2$

The equation for the air op. line can be determined using the point-slope equation for a line at point B. The coordinates are

$$T_B = 110°$$

$$h_B = a + b(96) + c(96)^2$$

$a = 2295$

$b = -63.85$

$c = 0.598$

$$h_B = 1676.6$$

at point C

$$T_c = 90°$$

$$h_c = a + b(70) + c(70)^2$$

$$h_c = 755.7$$

$$\text{slope} = B = \frac{1676.6 - 755.7}{110 - 90} = 46.045$$

$$\text{intercept} = A = -BT_B + h_B$$

$$A = -46.05(110) + 1676.6 = -3388.4$$

$$a - A = 2295 - (-3388.4) = 5683.4$$

$$b - B = -63.85 - 46.05 = -109.90$$

$$c = 0.598$$

$$q = 4(5683.4)(.598) - 109.90^2 = 1516.68$$

$$\sqrt{q} = 38.954$$

The "tower characteristic" becomes

$$\frac{KaV}{L} = \frac{2}{38.954}\left[\tan^{-1}\frac{2(.598)(110) - 109.90}{38.945} - \tan^{-1}\frac{2(.598)(90) - 109.90}{38.958}\right]$$

$$\frac{KaV}{L} = .029026$$

On the "cool" day

100°F, @ 50% RH is 83.5°F w.b.

3-18 (con't)

The enthalpy diagram can be drawn assuming L/V and the "tower characteristic" are unchanged.

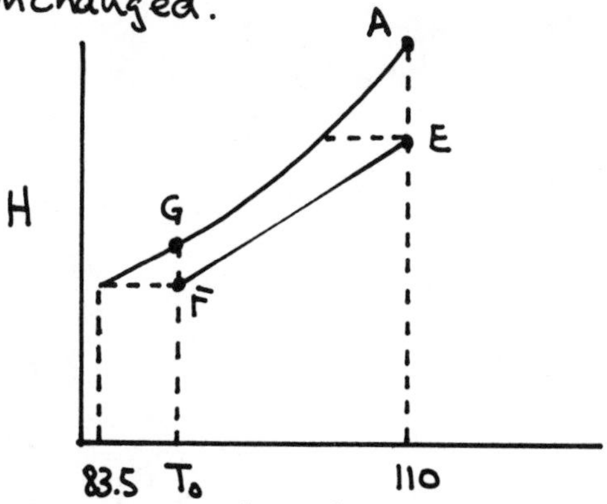

The air op line has same slope, $B = 46.05$ but different intercept, A_o. The cold water temperature is T_o.

The coordinates of point F

$$T_F = T_o$$

$$h_F = a + b(83.5) + c(83.5)^2$$

$$h_F = 1132.93$$

The opline, $A_o + BT$, is

$$B = 46.05$$

$$A_o = -BT_F + h_F = -46.05T_o + 1132.93$$

$$a - A_o = 1162.07 + 46.05T_o$$

$$b - B = -109.90$$

$$c = 0.598$$

$$q = 4(1162.07 + 46.05T_o)(.598) - 109.90^2$$

$$q = -9298.34 + 110.1516T_o$$

$$\frac{K_aV}{L} = 0.029026 =$$

$$\frac{2}{\sqrt{q}}\left[\tan^{-1}\frac{2(.598)(110) - 109.90}{\sqrt{q}} - \tan^{-1}\frac{2(.598)(T_o) - 109.90}{\sqrt{q}}\right]$$

trial & error for T_o

T_o	$\sqrt{q}$	$\frac{K_aV}{L} - 0.029026$
90	24.806	0.036157
97	37.322	−0.0946
94	32.4939	0.02385
94.2	32.831	0.00138
94.4	33.1648	0.000413
94.5	33.3042	−0.00005

close enough.

Cold water temperature increases from 90 to 94.5°F because wet bulb temperature increased when ambient temperature decreased.

3-19

$$2A \rightarrow P$$

At 90% conversion in ½ hour in a $\varepsilon = 0$ batch reactor with $C_{A_o} = 1$ mole/l

$$Kt = \frac{X_A}{C_{A_o}(1 - X_A)} = \frac{(.9)}{1(1 - .9)} = 9 \text{ liter/mole}$$

$$K = 18 \text{ hr}^{-1} \text{ liter mole}^{-1}$$

$\varepsilon \neq 0$

$$\varepsilon_A = \frac{V_{x=1} - V_{x=0}}{V_{x=0}} = \frac{1-2}{2} = -\frac{1}{2}$$

$$KC_{A_o}t = \frac{(1+\varepsilon_A)X_A}{(1-X_A)} + \varepsilon_A \ln(1-X_A)$$

$$18(1)t = \frac{(1-\frac{1}{2})(.9)}{(1-.9)} - \frac{1}{2}\ln(1-.9)$$

$$t = 0.31396 \text{ hr or } 18.8 \text{ min}$$

3-20

GRAPHICAL SOLUTION SAME AS PROBLEM 1-8. ANALYTICAL SOLUTION AVAILABLE ONLY IF EQUILIBRIUM AND TIE LINE DATA ARE AVAILABLE.

4-1

(a) $i = 7\%$

SCHEME I

$$EUAC_I = (25500 + 10000)(A/P, i, 10) - 2000(A/F, i, 10) + 10.5(24)(365)$$

$$EUAC_I = \$96890$$

SCHEME II

let
$A = 13000 + 6000 = 19000$
$B = 19500 + 5000 - 1500 = 23000$
$C = 2500 = 2500$
$L = 8(24)(365) = 70080$
$L1 = 0.5(24)(365) = 4380$

$$EUAC_{II} = [A/P, i, 10][A + B(P/F, i, 5) + L1(P/A, i, 5)] - C(A/F, i, 10) + L$$

$$EUAC_{II} = \$77496$$

COST PER TON

$$C_I = \frac{EUAC_I}{[P]\,24(365)} \quad \text{where } P = 50$$

$$C_I = \$0.2212/\text{ton}$$

$$C_{II} = \frac{EUAC_I}{[P]\,24(365)} \quad P = 20$$

$$C_{II} = \$0.442/\text{ton}$$

When a facility reaches capacity assume that another facility is needed to be built for higher total production

The cost to produce 1 ton/hr for scheme I & II is $11.06 and $8.86 respectively.

(b)

The cost/ton for each scheme at any production rate P becomes

$$C_I = \frac{\text{int}\left[1 + \frac{P-1}{50}\right](11.06)}{P}$$

$$C_{II} = \frac{\text{int}\left[1 + \frac{P-1}{20}\right]8.86}{P}$$

A plot of C_I & C_{II} vs P shows

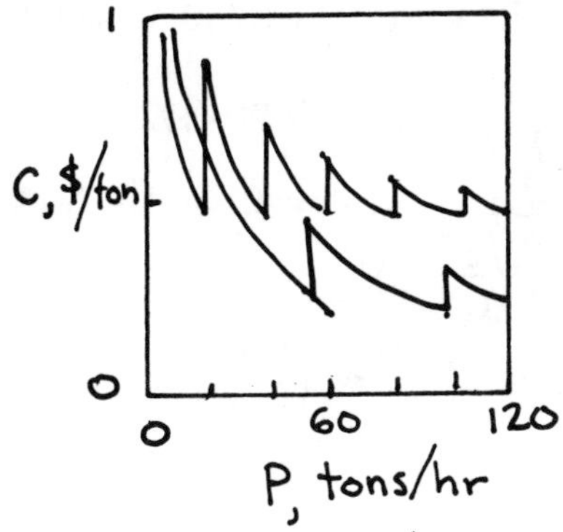

That SCHEME II has a lower cost/ton up to a production rate of 20, after 20 SCHEME I has a lower cost/ton. For rates > 20 tons/hr scheme I is better.

4-2

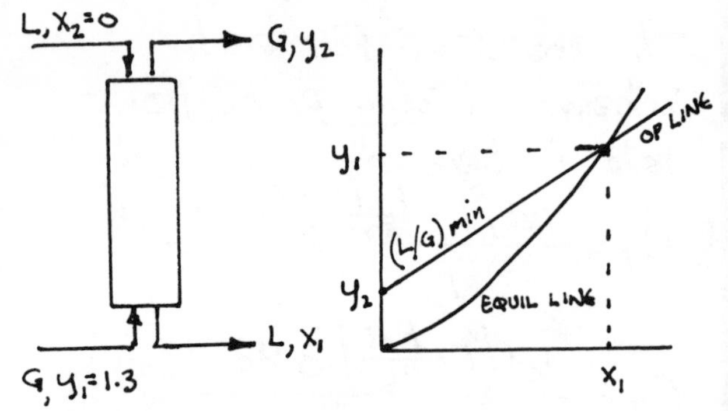

Assume: 100 lb moles/min

at 99% efficiency

gas out = 87 + 13(.01) = 87.13 lbmoles/min

$y_2 = \frac{.01(13)}{87.13} = 0.001492$

Op line intersects equil curve at y = 0.13

Op line $y_1 = y_2 + \left(\frac{L}{G}\right)_{min} X$

need equil data to find $\left(\frac{L}{G}\right)_{min}$

from Perry (5th)

C_A	P_A	y_A	X_A	m
5	665	.875	.01387	63.1
2.5	322	.4237	.00698	60.7
1.5	186	.2447	.00420	58.3
1.0	121	.1592	.00230	56.9
0.7	87	.1145	.00196	58.4
0.5	57	.0750	.00140	53.6

where $C_A \equiv$ lb SO_2/100 lb H_2O

$P_A \equiv$ mm Hg

$y_A \equiv P_A/760$

$X_A \equiv [C_A/64] / [5.56 + (\frac{C_A}{64})]$

where $5.56 = \frac{100 \text{ LB } H_2O}{18 \text{ LB/LB MOLE}}$

$m \equiv y_A/X_A$

$\bar{m} = 58.5$

$y_1 = y_2 + \left(\frac{L}{G}\right)_{min} X_1$

$X_1 = \frac{y_1}{m} = .13/58.5 = .00222$

$\therefore .13 = .001492 + \left(\frac{L}{G}\right)_{min}(.00222)$

$\left(\frac{L}{G}\right)_{min} = 57.9$ lbmoles H_2O/lbmole gas

4-2 (con't)

100 cfm: $\frac{100(1 \text{ atm})}{.73023(564°R)} = 0.2428 \frac{\text{lbmoles}}{\text{min}}$

$L = (57.9)(.2430)(18) = 253$ lb H_2O/min

4-3

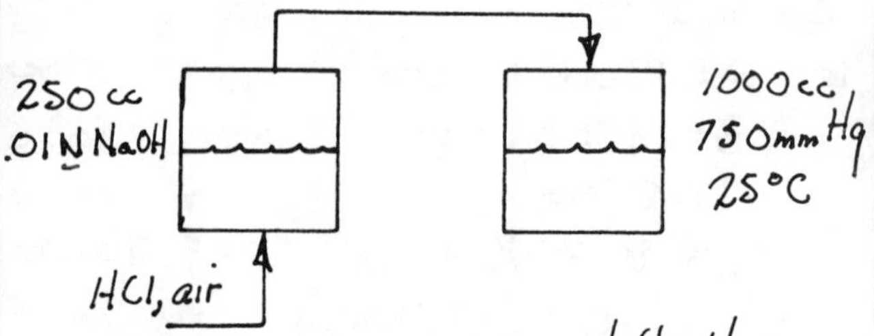

20 psia
20°C

$NaOH + HCl \rightarrow NaCl + H_2O$

moles:

NaOH(start) = 0.25(.01) = .0025 moles

NaOH(left) = .01749(.1) = .001749 moles

HCl scrubbed = .000751 moles

assume no water in air when volume measured in wet test meter

$\text{moles air} = \frac{PV}{RT} = \frac{(750/760)1000}{(82.06)(298)}$

$= .04036$ moles

assume water in air

$P_{H_2O} = 0.4594 \text{ psia} = 23.76 \text{ mm Hg}$

$P_{AIR} = 750 - 23.76 = 726.24 \text{ mm Hg}$

$\text{moles air} = \frac{PV}{RT} = \frac{(726.24/760)1000}{(82.06)(298)}$

$= .03907$ moles

WATER IN AIR

$y_{HCl} = \frac{.000751}{.000751 + .03907}$

$= .01886$

$Y_{HCl} = \frac{y_{HCl}(M_{HCl}/M_{air})}{1 + y_{HCl}\left(\frac{M_{HCl}}{M_{air}} - 1\right)}$

$Y = \frac{.01886(36.5/29)}{1 + .01886\left(\frac{36.5}{29} - 1\right)}$

$Y = 0.02362$

NO WATER IN AIR

$y = \frac{.000751}{.000751 + .04036}$

$= .01827$

$Y = \frac{.01827(36.5/29)}{1 + .01827\left(\frac{36.5}{29} - 1\right)}$

$Y = 0.02289$

4-4

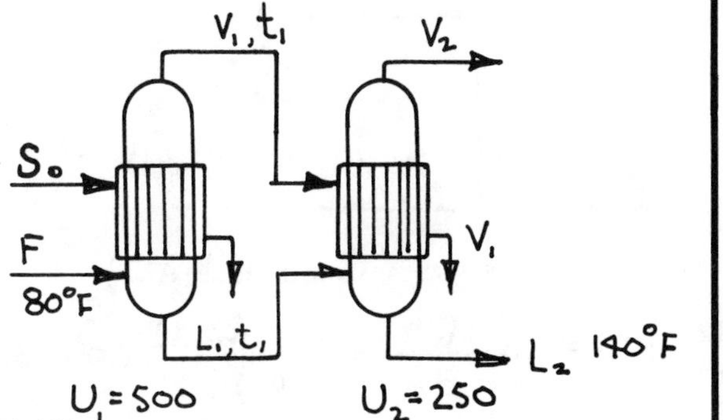

ASSUME:

① $A_1 = A_2$

② $V_1 = V_2$

③ $C_p = 1$

④ feed 100 lb/hr

⑤ no heat loss, neglect BPE, condensate not subcooled

METHOD:

Guess Δt in each effect, do heat balance around each effect, calculate steam rate, calculate areas, if not equal use average area to find new Δt for each area. Repeat until areas EQUAL.

$F = V_1 + V_2 + L_2$

$0.15F = 0(V_1 + V_2) + 0.6L_2$; $L_2 = 25$ lb/hr

$V_1 = V_2 = 37.5$ lb/hr

$\Delta t_{overall} = t_{steam} - t_{product} = 228 - 140 = 88°F$

$Q_1 = V_1\lambda_1 + FC_p(t_1 - t_{feed})$

$Q_2 = V_2\lambda_2 + L_1 C_p(t_{PROD} - t_1)$

$A_i = Q_i/U_i\Delta t_i$; $S_o = Q/\lambda_o$; $\lambda_o = 960.1$

For 1st iteration, assume

$\Delta t \propto 1/u$

For next iteration, set

$\Delta t_1 = Q_1/U_1 A_{average}$

initial guess: $\dfrac{\Delta t_1}{\Delta t_2} = \dfrac{U_2}{U_1} = \dfrac{250}{500}$

$\Delta t_1 + \Delta t_2 = 88$

$\Delta t_1 \approx 28 \quad \Delta t_2 \cong 60$

4-4 (con't)

$L_1 = V_2 + L_2$
$= 37.5 + 25$
$= 62.5$ lb/hr.

	1st iteration	2nd iteration	3rd iteration
Δt_1	28	35	36
Δt_2	60	53	52
λ_1	977.9	982.2	982.8
λ_2	1014.0	1014	1014
Q_1	48671	48132.5	48055
Q_2	34275	34712.5	3477.5
A_1	3.477	2.75	2.670
A_2	2.285	2.62	2.6715
A_{avg}	2.881	2.685	2.673
S_o	50.7	50.1	50.05
t_1	200	193	192
t_2	140	140	140

final conditions:

i	Δt_i	t_i °F	p_i, psia
0	-	288	20
1	36	192	9.75
2	52	140	2.89

4-5

An enthalpy-concentration diagram for H_2SO_4 is need. Refer to Perry 5th Ed pg 3-204.

Enthalpy 98% H_2SO_4 @ 70°F = 0 Btu/lb
H_2O @ 70°F = 38 Btu/lb

Mixture: $\% = \dfrac{\rho_a(50)(.98)}{\rho_a 50 + \rho_w(500)} \times 100$

$\rho_a = 1.835 \quad \rho_w = .9980$ g/cc

$\% H_2SO_4 = \dfrac{1.836(50).98}{1.836(50) + .998(500)} \times 100$

$= 15.2\%\ H_2SO_4$

enthalpy of 15.2% H_2SO_4 =

$\dfrac{0 + \rho_w(500)\,38}{\rho_A 50 + \rho_w 500} = 32.1$ Btu/lb

4-5 (con't)

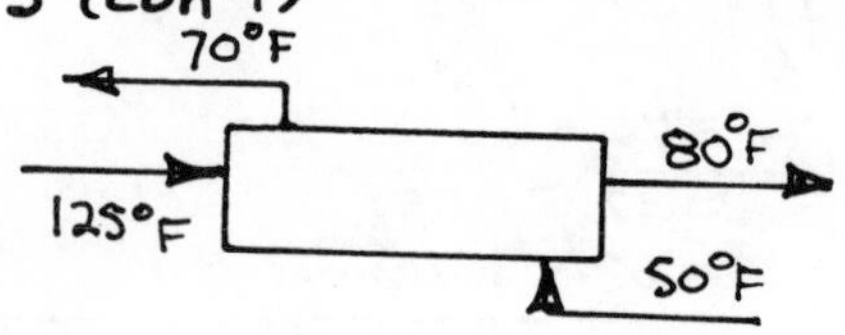

Resulting enthalpy calculated lies on line between starting composition on h-concentration diagram

from diagram $t \cong 125°F$

$\Delta t_1 = 125-70 = 55°F$

$\Delta t_2 = 80-50 = 30°F$

$\bar{\rho}_{H_2SO_4} = \frac{1.0861+1.099}{2} = 1.0930$

$\Delta t_{lm} = \frac{55-30}{\ln \frac{55}{30}} = 41.2°F$

flow rate $= \frac{550}{30}(8.345)(1.0930) = 167.2 \frac{lb}{min}$

$Q = mc_p\Delta t$ $\quad C_p = 0.877$ (Perry 5th)

$Q = 167.2(.878)(125-80) = 6606.1 \frac{Btu}{min}$

$= 396364 \frac{Btu}{hr}$

$Q = UA\Delta t_{lm} = 500\,A\,(41.2) = 396364$

$A = 19.24 ft^2$

4-6 $N_2 + C_2H_2 \rightleftharpoons 2HCN$; $G = H - TS$

$\Delta H = 30800(2) - 53900 - 0 = 7700 \frac{cal}{mole}$

$T\Delta S = (298)[48.25(2) - 47.5 - 45.79]$

$= 956.58 \frac{cal}{mole}$

$\Delta G = 7700 - 956.58 = 6743.42 \frac{cal}{mole}$

$\Delta G = -RT\ln K = -1.987(298)\ln K = 6743.42$

$K = 1.133 \times 10^{-5}$

reaction is doubtful. EQUILIBRIUM does not favor the production of hydrogen cyanide in this manner. Only when $\Delta G < 0$ is reaction promising. Thermodynamics does not predict kinetics which if forward rate > reverse, the reaction may be possible

4-7

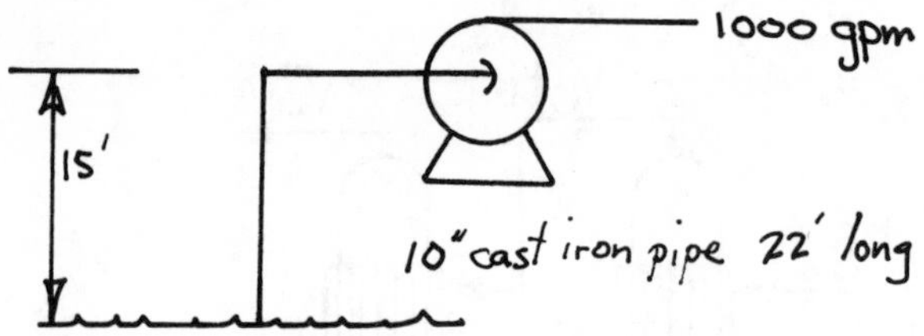

$\varepsilon = 0.00085$ (Crane, Table A23)

$N_{Re} = 50.6 \frac{Q\rho}{d\mu} = \frac{1000(62.4)}{(10.02)(1)} 50.6 = 315113$

$f = 0.019$ Crane TABLE A24

$NPSH = (h_p - h_{vp}) - h_z - h_f$

$h_p = (29.9)(1.133) = 33.8767 ft\ H_2O$

$h_{vp} = 0.78\ ft\ H_2O$ @ 68°F

$h_z = -15\ ft\ H_2O$

$h_f = \left[f\frac{L}{D} + K_e + K_v + K_f\right]\frac{v^2}{2g}$

fitting = 0.75

valve = 0.17

entrance = .00

$NPSH = 33.867 - .78 - 15 - \left[.019\frac{22}{10/12} + 0 + .17 + .75\right]\frac{4.08^2}{2(32.2)}$

$NPSH = 17.72\ ft\ H_2O$

4-8

Variable volume PFR 2nd order:

$k\tau = \frac{1}{C_{A_0}}\left[2\varepsilon_A(1+\varepsilon_A)\ln(1-X_A) + \varepsilon_A^2 X_A + (\varepsilon_A+1)^2\frac{X_A}{1-X_A}\right]$

$V_{x=0} = 20 + 80 = 100$ vols $\quad V_{x=1} = 20 + 40 = 60$ vols

$\varepsilon_A = \frac{60-100}{100} = -0.4$

molar volume $= 22.4 \frac{l}{mole}\left(\frac{599°K}{273°K}\right)\frac{1}{10} = 4.915 \frac{l}{mole}$

$C_{A_0} = 0.8/4.915 = 0.1628$ moles/liter

$X_A = .9$

$\tau = \frac{1}{(.85)(.1628)}\left[2(-.4)(.6)\ln(.1) + .4^2(.9) + .6^2\frac{.9}{.1}\right]$

$\tau = 32.44 min$

4-8 (con't)

$$F_{A_0} = 20000 \frac{\text{lb}}{\text{day}} \frac{1}{24(60)} \times \frac{1 \text{ lbmole}}{54 \text{ lb}} \times 454 \frac{\text{gmole}}{\text{lbmole}}$$

$F_{A_0} = 116.77$ gmole/min

$$F_{A_0}/C_{A_0} = v_0 = 116.77/.1628 = 717.26 \frac{\ell}{\text{min}}$$

$$V = v_0 \tau = 717.26(32.44) = 2.327 \times 10^4 \ \ell$$

4-9

Regression using linear transformation

Linearize Arrhenius Equation

$$k = k_0 e^{-E/RT}$$

to

$$Y = mX + b$$

where

$$Y = \ln K$$

$$b = \ln K_0$$

$$m = -\frac{E}{R}$$

$$x = \frac{1}{T}$$

$$m = \frac{\sum xY - \sum x \sum Y/n}{\sum x^2 - (\sum x)^2/n}$$

$$b = \frac{\sum y - m \sum x}{n}$$

t	T	$K \times 10^5$	Y	xY	$X^2 \times 10^5$	X
0	273	1.2	-11.33	-.0415	1.342	.00366
15	288	8.34	-9.39	-.0326	1.206	.00347
30	303	46.5	-7.67	-.0253	1.089	.00330
50	323	360	-5.63	-.0174	0.9585	.0031
Σ			-34.02	-.1168	4.5951	.01353

4-9 (con't)

$$m = \frac{-.1168 - .01353(-34.02)/4}{4.5951 \times 10^{-5} - (.01353)^2/4}$$

$m = -9298.$

$b = 22.945$

$E = -mR = -(-9298)(1.987) = 18475$ cal/mole

$K_0 = 9.223 \times 10^9$

$$K = 9.223 \times 10^9 \, e^{-9298/T}$$

4-10 same as 1-14

4-11

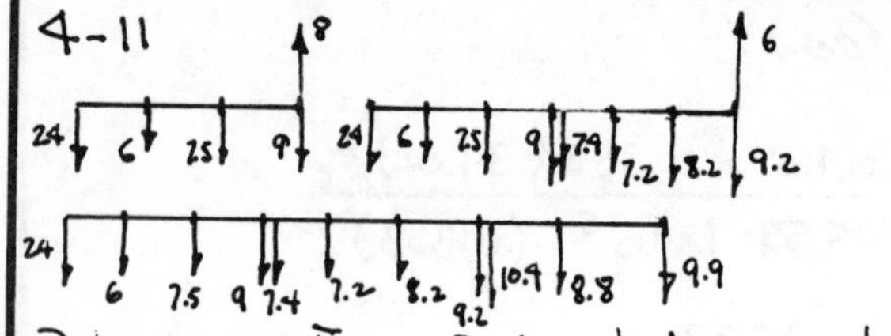

PROCEDURE: FIND PW cost & CONVERT to EUAC [+ ≡ cost]

PW°_j ≡ Present worth for j years without salvage

$I = 24000 \quad i = 10\%$

$C_1 = 6000 \quad C_2 = 7500 \quad C_3 = 9000$

$O_3 = 7400 \quad S_3 = 8000$

$C_4 = 7200 \quad C_5 = 8200 \quad C_6 = 9200$

$O_6 = 10400 \quad S_6 = 6000$

$C_7 = 8800 \quad C_8 = 9900 \quad S_8 = 0$

$PW^{\circ}_3 = I + C_1(P/A,i,3) + 1500(P/G,i,3)$

$PW^{\circ}_3 = \$42415$

$PW_3 = PW^{\circ}_3 - S_3(P/F,i,3)$

$PW_3 = \$36404$

$EUAC_3 = PW_3(A/P,i,3)$

$EUAC_3 = \$14639/\text{year}$

$PW^{\circ}_6 = PW^{\circ}_3 + O_3(P/F,i,3) + C_4(P/F,i,4) + C_5(P/F,i,5) + C_6(P/F,i,6)$

$PW^{\circ}_6 = 63177$

$PW_6 = PW^{\circ}_6 - S_6(P/F,i,6)$

$PW_6 = 59790$

$EUAC_6 = PW_6(A/P,i,6)$

$EUAC_6 = \$13728/\text{year}$

$PW_8 = PW^{\circ}_6 + O_6(P/F,i,6) + C_7(P/F,i,7) + C_8(P/F,i,8)$

$PW_8 = \$78182 \qquad EUAC_8 = PW_8(A/P,i,8)$

$EUAC_8 = \$14655/\text{year}$

BULLDOZER SHOULD BE SALVAGED AFTER 6 YEARS SINCE THE EUAC IS MINIMIZED BY THAT OPTION.

4-12

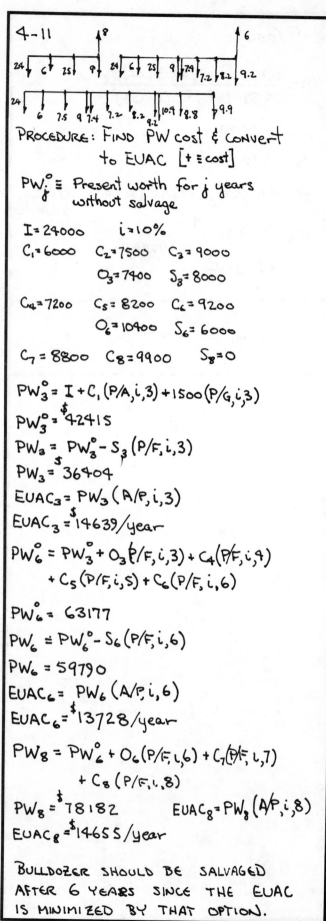

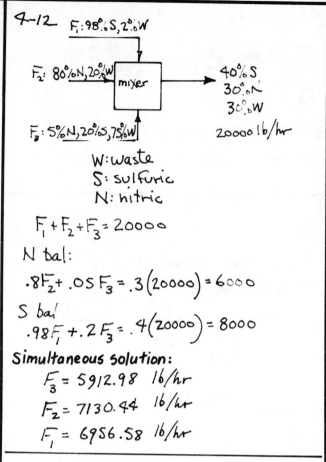

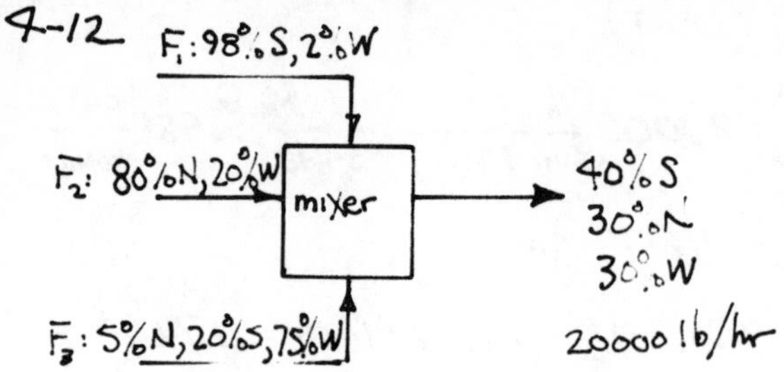

W: waste
S: sulfuric
N: nitric

$F_1 + F_2 + F_3 = 20000$

N bal:

$.8F_2 + .05F_3 = .3(20000) = 6000$

S bal

$.98F_1 + .2F_3 = .4(20000) = 8000$

Simultaneous solution:

$F_3 = 5912.98$ lb/hr

$F_2 = 7130.44$ lb/hr

$F_1 = 6956.58$ lb/hr

4-13

$C_A = C_{A_0}/(K\tau+1)$

at 99% steady state

$C_A = .99\,C_{A_0}/(K\tau+1)$

unsteady state:

input − output = accumulation + disap Rx

$v_0 C_{A_0} - v_0 C_A = V\dfrac{dC_A}{dt} + (-r_A V)$

$-r_A = KC_A \qquad \tau = V/v_0$

$\therefore\; C_{A_0} - C_A - K\tau C_A = \tau\dfrac{dC_A}{dt}$

$$\int_0^t dt = \tau\int_{C_{A_0}}^{C_A}\frac{dC_A}{C_{A_0} - C_A(K\tau+1)}$$

$$t = \frac{\tau}{K\tau+1}\ln\frac{C_{A_0}K\tau}{C_{A_0} - C_A(K\tau+1)} \qquad \text{plug in } C_A = \frac{.99C_{A_0}}{K\tau+1}$$

$$t = \frac{\tau}{K\tau+1}\ln\frac{C_{A_0}K\tau}{C_{A_0} - .99C_{A_0}} = \frac{\tau}{K\tau+1}\ln(100K\tau)$$

$K = .42\ \text{min}^{-1} \quad \tau = 2\ \text{min}$

$$t = \frac{2}{.42(2)+1}\ln[10(.42)(2)] = 4.82\ \text{minutes}$$

4-14

Same as 3-2

4-15

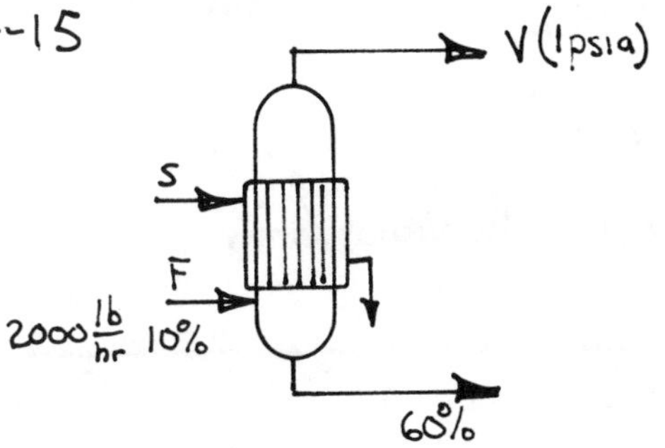

assume $C_p = 1$

neglect BPE feed at 70°

$F = V + P = 2000$

$.1F = .6P$

$P = 333$ lb/hr

$V = 1667$ lb/hr

$Q = 2000\, c_p (t_{evap} - t_{feed}) + 1667 \lambda_{evap}$

steam tables $t_{sat} = 102°F$ @ 1psia

$\lambda = 1036$ Btu/lb

$Q = 2000(1)(102-70) + 1667(1036)$

$Q = 1.79 \times 10^6$ Btu/hr

$t_{steam} = 338°F$ @ 114.7 psia

$A = Q/U\Delta t = 1.79 \times 10^6 / (60[338-102])$

$A = 126$ ft²

use 400 ft² evap.

4-16

Present freight cost:

$\frac{10^6 \text{ lb/yr}}{.5} \times .028 \frac{\$}{lb} = \$56000/yr$

Concentrated freight cost

$\frac{10^6}{.9} \times .028 = \$3111/yr$

Steam cost

$\left[\frac{10^6}{.5} - \frac{10^6}{.9}\right]\left(1.2 \frac{lb\ steam}{lb\ H_2O}\right)\left(.003 \frac{\$}{lb}\right) = \$3200$

\$ avail for conc. = 56000 - 3111 - 3200 = \$21689/yr

$.3(FCI) + .2(FCI) = 21689$; $(FCI) = \$43378$

(ROR) (COSTS)

4-17

(a) **Constant rate time**

$$\Theta_c = \frac{-W_s}{A r_c}(X_c - X_i)$$

Falling rate time

$$\Theta_r = \frac{-W_s X_c}{A r_c} \ln \frac{X_f}{X_c}$$

W_s = weight of solids, lb

A = drying area, ft²

r_c = constant rate drying lb/hr-ft²

X_c = critical moisture

X_i = initial moisture

X_f = final moisture

Empirical relationships

$$r_c = \frac{h}{\lambda}(t_v - t_s)$$

h = heat transfer coeff

λ = heat of evap

t_v = vapor temp

t_s = solid surface temp (= wet bulb temp)

$$h = 0.0128 (G_v)^{0.8}$$

G_v = gas rate lb/hr-ft²

$\therefore h = 0.0128 (1200)^{.8} = 3.72$ Btu/hr-ft²-°F

$\lambda = 1008$ Btu/lb at 150°

$t_i = 78.5°F$ (wet bulb for .005 = H)

$r_c = \frac{3.72}{1008}(150 - 78.5) = 0.264 \frac{lb}{hr-ft^2}$

$W_s = 90 \times \frac{1 \text{ inch}}{12} = 7.5 \frac{lb}{ft^2}$ or 7.5 lb based on $A = 1$ ft²

$\Theta_{TOTAL} = \Theta_c + \Theta_r = \frac{-7.5}{(1)(.264)}(0.1 - 0.5)$

$+ \frac{-7.5(.1)}{(1).264} \ln \frac{.02}{.1}$

$\Theta_{TOTAL} = 11.364 + 4.572 = 15.936$ hr

4-17 (con't)

(b) at 200°F

$\lambda = 977.9$

$t_i = 90°F\ (\sim 89.5°F)$

$r_c = \frac{3.72}{977.9}(200-89.5) = 420\ \frac{lb}{hr\text{-}ft^2}$

$\Theta_T = \frac{-7.5}{(1)(420)}\left[(0.1-0.5) + .1\ \ln\frac{.02}{.1}\right]$

$= 10.0$ hr

4-18

(a) $Q = \sqrt{\frac{\Delta P\ (62.4)}{\rho}}\quad C_v$

$\Delta P = 1000$

$\rho = 39.00$ from steam tables ($\bar{V} = .0257$)

$C_v = 400$

$Q = 16018$ gpm

$q = 0.1337Q$ ft³/min

$q = 2142$ ft³/min

$W = q\rho$ lb/min

$W = 83333$ lb/min

(b) @ 2000 psia

$V_L = 1/\rho$

$V_L = 0.257$ ft³/lb

$h_L = 671.7$ Btu/lb steam tables

@ 1000 psia

$h1_L = 542.41$ Btu/lb

$h1_v = 1191.8$ Btu/lb

$W_L = W\frac{h_L - h1_v}{h1_L - h1_v}$

4-18 (con't)

$W_L = 66742$ lb liquid/min

$W_L = W - W_L$

$W_L = 16591$ lb vapor/min

4-19

$77°C = 350°K$

$\ln P_b = 17.61922 - \frac{3880}{350}$; $P_b = 687.8$ mm

$\ln P_c = 17.40484 - \frac{3811.61}{350}$; $P_c = 674.9$ mm

at azeotrope:

$\gamma_b = 687.8/760 = 0.905$

(a) $\gamma_c = 674.9/760 = 0.888$

$A_{12} = \ln\gamma_1\left(1 + \frac{X_2 \ln\gamma_2}{X_1 \ln\gamma_1}\right)^2$

$X_b = 0.55 = y_b$

$A_b = \ln .905\left(1 + \frac{.45 \ln .888}{.55 \ln .905}\right)^2 = -0.3888$

$A_c = \ln .888\left(1 + \frac{.55 \ln .905}{.45 \ln .888}\right)^2 = -0.4881$

(b) at $X_b = 0.2$

$\ln\gamma_b = \frac{-0.3888}{\left[1 + \frac{-0.3888}{-0.4881}\left(\frac{.2}{.8}\right)\right]^2}$

$\gamma_b = 0.7631$

similarly,

$\gamma_c = 0.9866$

$y_b = \frac{\gamma_b X_b P_b^\circ}{P} = (.7631)(.2)P_b^\circ/760$

$y_b = 2.008\times10^{-4}\ P_b^\circ$

$y_c = 10.39\times10^{-4}\ P_c^\circ$

4-19 (con't)

trial & error to find temperature:

t	P_b°	P_c°	y_b	y_c	Σy
77	687.8	674.9	.1381	.7012	.8393
87	935.8	913.3	.1879	.9489	1.1368
82	804.0	786.7	.1615	.8174	.9789
82.7	821.5	803.6	.1650	.8349	.9999

$t = 82.7\,^\circ C$

4-20

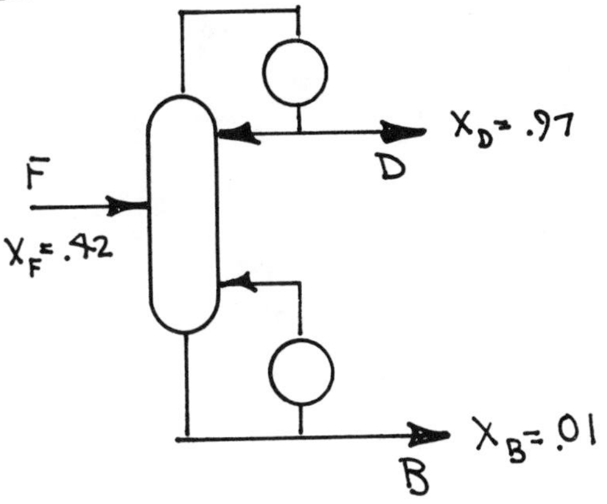

$F = 1 = D + B$

$.47F = .97D + .01B$

$$D = F\frac{X_F - X_B}{X_D - X_B} = (1)\frac{.42 - .01}{.97 - .01}$$

$D = .427 \qquad B = .573$

Rectifying op line ("R" line)

$$y = \frac{R}{R+1}x + \frac{1}{R+1}X_D$$

$$y = \frac{2.5}{3.5}x + \frac{1}{3.5}\,.97$$

$y = .7143x + .2771$

Q-line

$X_F = .47$

4-20 (con't)

Strip line ("S" line)

above feed plate:

$L = RD = 2.5(.427) = 1.0675$

below feed plate

$L = 1 + 1.0675 = 2.0675$

op line:

$$y = \frac{L}{L-B}x - \frac{Bx_B}{L-B}$$

$$y = \frac{2.0675}{2.0675 - .573} - \frac{.573(.01)}{2.0675 - .573}$$

$y = 1.383x - 0.00383$

Coordinates of end points of op lines

	x	y
R-line	.97	.97
R-line	.42	.577
S-line	.42	.577
S-line	.01	.01

For equil line assume Raoult's Law

$y_i = x_i P_i^\circ / P$

Since $\log_{10} P_i = A - \frac{B}{C+t}$ $\quad t\,^\circ C$

and feed at boiling

4-20 (con't)

find temp where

$760 = X_1 P_1^o + X_2 P_2^o = \Sigma$

$t\,°C$	Σ
80	252.5
100	483.1
110	648.9
115	747.1
115.62	760.0

$X_1 = .42$

$X_2 = .58$

$P^o_{HEPTANE} = 1226.69$

$P^o_{ETHYLBENZENE} = 422.15$

$\alpha = 2.906 = P^o_H / P^o_E$

FROM PLOT: 11+ theoretical stages req'd

PROBLEM 4−20

McCabe−Thiele Plot

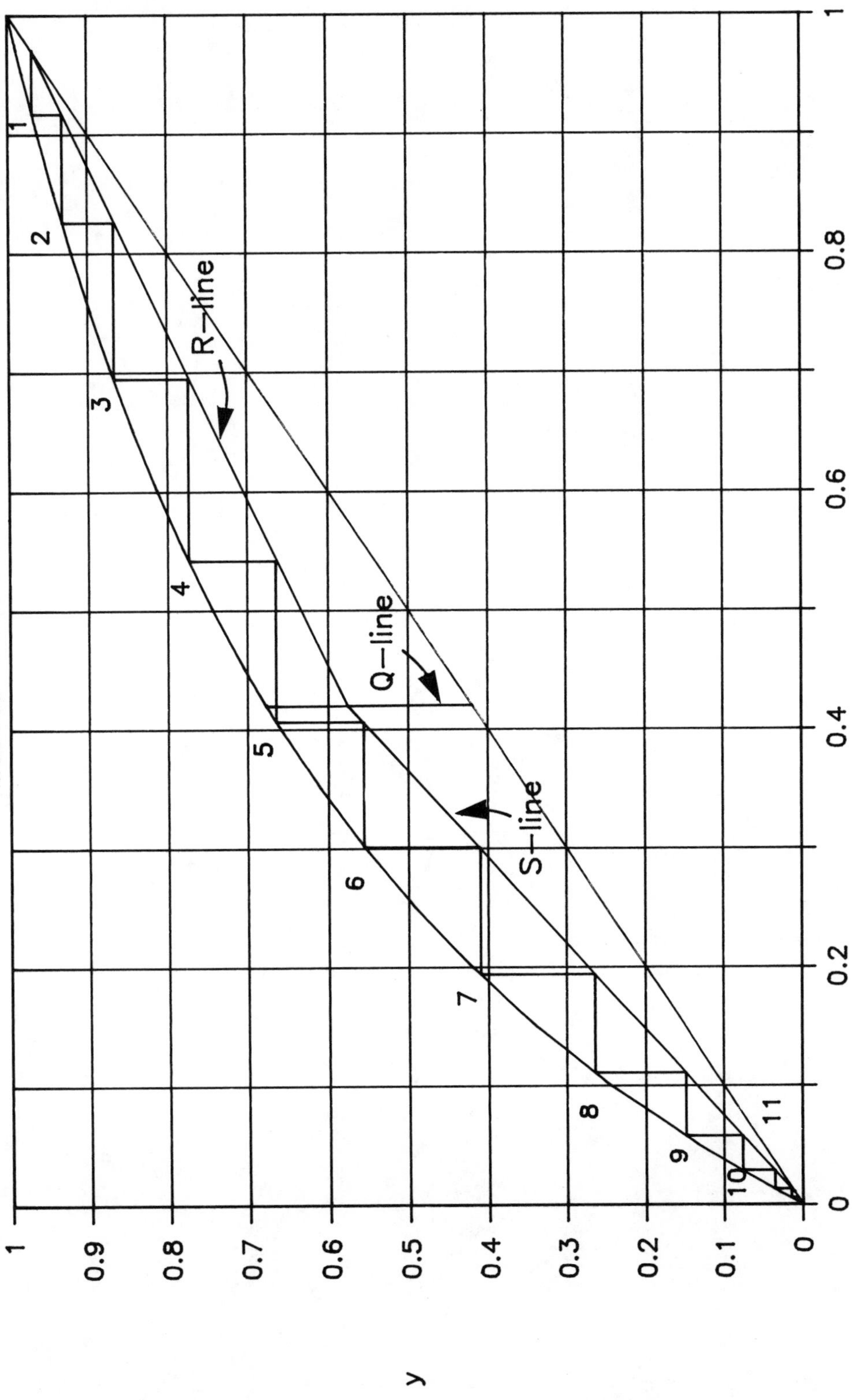

5-1

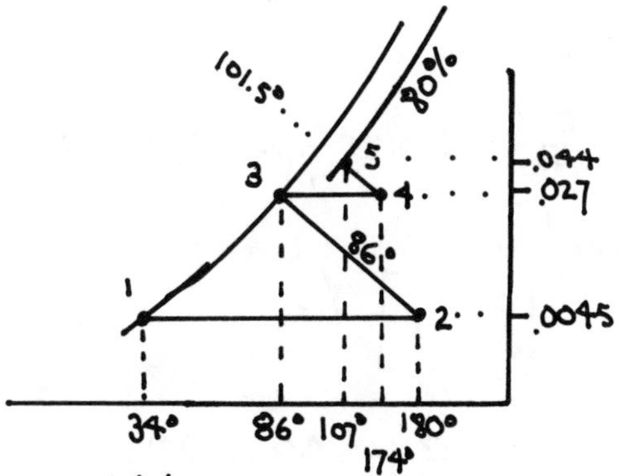

(a) dryer 1 (state 2-3): $H_2O = .027 - .0045 = .0225 \frac{lb\,H_2O}{lb\,d.a.}$ evap

· dryer 2 (state 4-5): $(.044 - .027) = .0170 \frac{lb\,H_2O}{lb\,d.a.}$ evap

(b) from humid heat vs humidity heat at .0045 humidity

$C_p = 0.242$ Btu/lbd.a.°F

heater 1 duty $= C_p \Delta t = 0.242(180-34)$
$= 35.332$ Btu/lbd.a.

heater 2 duty: $C_p = .2525$ Btu/lb°F

duty $= .2525(174-86) = 22.22$ Btu/lb

(c) dryer 1 exit air is saturated

Sat vol = 13.7 ft³/lb d.a. at 86°F

dryer 2 exit at 80% RH at 107°

$$\text{sp. vol} = \text{dry sp. Vol} + \frac{\%h}{100}(\text{sat Vol} - \text{dry Vol})$$

$$= 14.25 + 0.8(15.55 - 14.25) = 15.29\ \frac{ft^3}{lb}$$

5-2

The tower will perform adequately only if the tower does not flood at new rate

This is a problem of calculating flooding conditions using the following conditions and assumptions.

5-2 (con't)

(1) packing is dumped, not stacked

(2) Tower operates reasonably far from flooding at old rates.

$$\Delta P = 0.5\text{ psi} \times 2.307\left(\frac{12}{14}\right)$$
$$= 13.84 \text{ in } H_2O/14\text{ ft}$$
$$= 0.998 \text{ in } H_2O/\text{ft packing}$$

$$L_{OLD} = \frac{440\text{ gpm} \times 60 \times 8.341}{\left(\frac{64}{12}\right)^2 \frac{\pi}{4}} = 9869\ \frac{lb}{hr\text{-}ft^2}$$

from Perry 5th pg 18-27

for 2" Raschig rings

$$G/\phi \cong 1000;\ \phi = \sqrt{\rho_v/.075} \cong 1$$

$G_{OLD} \cong 1000$ lb/hr-ft² (gas flow)

to determine new condition (if flooding) use Fig 18-51 Perry 4th pg 18-30

old:

$$\Delta P = 1;\ X_{OLD} = \frac{L}{G}\sqrt{\frac{\rho_v}{\rho_L}} = \frac{9869}{1000}\sqrt{\frac{.075}{62.4}}$$

$$X_{OLD} = .3421$$

$$Y_{OLD} = \frac{G a \psi^2 \mu^2}{\rho_v \varepsilon^3 \rho_L g_c}\left(\frac{1}{3600}\right)^2 = .035$$

new:

$\Delta P = ?$

$X_{new} = X_{OLD}$

$Y_{NEW} = Y_{OLD}(1.6)^2 = 0.0896$ (above flooding)

new conditions will flood tower. Tower will not perform its function.

5-3

θ	$\Delta\theta$	V	ΔV	$\Delta\theta/\Delta V$	$\bar{V}$
0	–	0	–	–	–
2.1	2.1	.32	.32	6.5625	.16
12	9.9	.97	.65	15.2308	.485

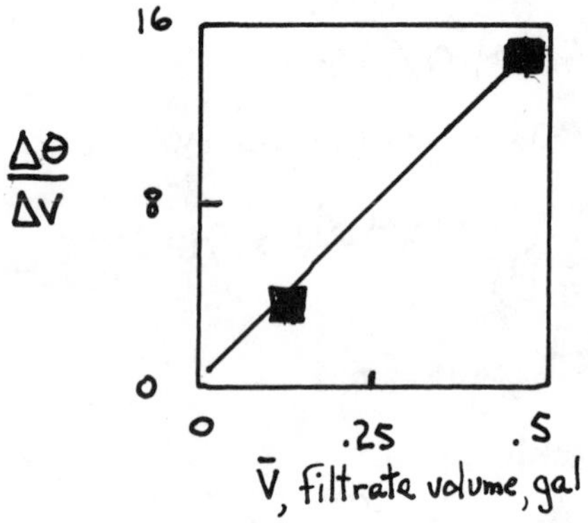

slope = K_p

intercept = C

$$\frac{d\theta}{dV} = \bar{V}K_p + C$$

$$K_p = \frac{15.2308 - 6.5625}{.485 - .160} = 26.67$$

$$C = 15.2308 - 26.67(.485)$$

$$C = 2.296$$

since $K_p = \frac{\omega\alpha\mu}{A^2\Delta P}$; $C = \frac{r\mu}{A\Delta P}$

$$K_{p\,plant} = K_{p\,lab}\frac{A^2_{lab}}{A^2_{plant}}\frac{\Delta P_{lab}}{\Delta P_{plant}}$$

$$K_{p\,plant} = 26.67\left(\frac{.5}{3}\right)^2\left(\frac{11}{30}\right) = 0.2716$$

$$C_{plant} = C_{lab}\frac{A_{lab}}{A_{plant}}\frac{\Delta P_{lab}}{\Delta P_{plant}}$$

$$C_{plant} = 2.296\left(\frac{.5}{3}\right)\left(\frac{11}{30}\right) = 0.1403$$

Plant:

$$\frac{d\theta}{dV} = 0.2716\bar{V} + 0.1403$$

@ $\bar{V} = 0$; $\frac{d\theta}{dV} = 0.1403$; $\frac{dV}{d\theta} = 7.127$ gpm

5-3 (con't)

final rate:

$$\frac{dV}{d\theta} = 0.7127 \text{ gpm} ; \frac{d\theta}{dV} = 1.4031$$

final volume

$$V = \left[\frac{d\theta}{dV} - C\right] / K_p$$

$$V = (1.4031 - .1403)/.2716 = 4.649 \text{ gal}$$

total time:

$$\theta = \int_0^V (K_pV + C)\,dV = \left[\frac{K_p\bar{V}^2}{2} + C\bar{V}\right]_0^V$$

$$\theta = \frac{.2716(4.649)^2}{2} + 0.1403(4.649)$$

θ = 3.59 minutes for each leaf

total cycle time

$$\theta_T = 3.59 + 8 = 11.59 \text{ min}$$

throughput

$$(4.649)/(11.59) = 0.401 \text{ gpm}$$

for press:

$$(0.401)(20) = 8.02 \text{ gpm}$$

not enough capacity Need higher pressure or more leaves

5-4

$$y_i = Kx_i$$

$$\Sigma y_i = 1 = K_1x_1 + K_2x_2 + K_3x_3$$

$$1 = .02\frac{100}{P}\frac{1}{3} + .03\frac{100}{P}\frac{1}{3} + .05\frac{100}{P}\frac{1}{3}$$

$$P = 3.33$$

$$\Delta P = 5 - 3.33 = 1.667 \text{ atm}$$

5-5

For 2 CSTR's in series, where $\varepsilon_A = 0$ and equal volumes

$$(k\tau + 1)^2 = \frac{C_{A0}}{C_{A2}} = \frac{1}{1 - X_A}$$

(a) at $X_A = 0.75$ $\quad (k\tau+1)^2 = \frac{1}{1-.75} = 4$

$$k\tau + 1 = 2 \ ; \ k\tau = 1$$

at 25°C $\quad \ln k = 8 - \frac{2928}{298}$

$$k = 0.16114$$

$$\tau = 1/0.16114 = 6.206 \text{ min}$$

$$V = \tau v_0 = 6.206(50) = 310.3 \text{ gals}$$

(b) at $X_A = 0.9$

$$(k\tau+1)^2 = 10 \ ; \ k\tau = 2.1623$$

$$k = 2.1623/6.206 = 0.3484$$

solve for T

$$T = B/(A - \ln k) = 2928/(8 - \ln .03484)$$

$$= 323.4°K$$

(c) at 90% conversion

$$k = 0.16114$$

$$\tau = 2.1623/0.16114 = 13.419 \text{ min}$$

$$V = \tau v_0 = 13.419(50) = 670.9 \text{ gal}$$

5-6

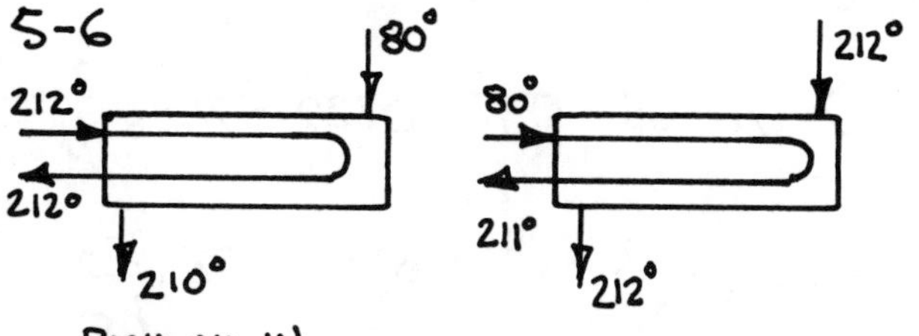

RICH OIL IN SHELL
I

LEAN OIL IN TUBES
II

I: $Q = mc_p\Delta t = 11000(.4)(210-80)$

$$= 572000 \text{ Btu/hr}$$

$$U_1 = Q/(A\,\Delta t_{lm})$$

$$\Delta t_{lm} = 132-2/\ln\left(\tfrac{132}{2}\right) = 31.03$$

5-6 (con't)

$$U_1 = 572000/(100(31.03))$$

$$= 184.3 \text{ Btu/hr ft}^2\text{°F}$$

II: $Q = 10000(.4)(211-80) = 524000 \ \frac{\text{Btu}}{\text{hr}}$

$$\Delta t_{lm} = \frac{132-1}{\ln 132} = 26.83°$$

$$Q = U_2 A\,\Delta t_{lm}$$

$$U_2 = 524000/(100(26.83))$$

$$U_2 = 195.3 \text{ Btu/hr ft}^2\text{°F}$$

Since U_1 & U_2 are nearly equal and steam condensing yields $h = 100$ to 300 Btu/hr ft²°F

using $h_{steam} = 200$

$$\frac{1}{U_1} = \frac{1}{h_{oil\ inside}} + \frac{1}{200} = \frac{1}{185}$$

$$\frac{1}{U_2} = \frac{1}{h_o} + \frac{1}{200} = \frac{1}{195}$$

$$h_i = 2348 \text{ Btu/hr ft}^2\text{°F}$$

$$h_o = 8311 \text{ Btu/hr ft}^2\text{°F}$$

With lean oil in tubes rich oil in shell so that:

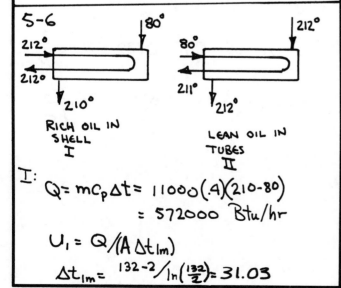

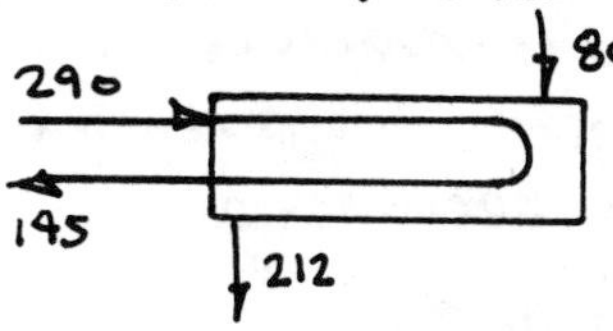

$$\Delta t_{lm} = \frac{210-67}{\ln \frac{210}{67}} = 125°$$

$$\frac{1}{U} = \frac{1}{h_i} + \frac{1}{h_o} = \frac{1}{2348} + \frac{1}{8311}$$

$$U = 1831 \text{ Btu/hr ft}^2\text{°F}$$

$$Q = 11000(.4)(212-80) = 580800 \ \frac{\text{Btu}}{\text{hr}}$$

$$A = Q/U\Delta t = 580800/[(1831)(125)]$$

$A = 2.5 \text{ ft}^2$ ∴ exchanger is adequate

5-7

(a) For CSTR, 1st order, $\varepsilon_A = 0$:

$$K\tau = \frac{X_A}{1-X_A}$$

BATCH:

$$kt = -\ln(1-X_A)$$

for constant density systems $\tau = \bar{t}$

so that

$$K\tau = Kt = \left\{\frac{X_A}{1-X_A}\right\}_{CSTR} = \left\{\ln(1-X_A)\right\}_{BATCH}$$

for CSTR $X_A = .75$ $K\tau = \frac{.75}{1-.75}$

$$K\tau = 3$$

$\therefore$ $kt = 3 = -\ln(1-X_A)$; $X_A = 0.95$ for batch

(b) assuming 100 lb mole feed

CSTR production:

$$R_{CSTR} = 100(.75) = 75 \text{ moles/hr}$$

BATCH

$$R_{BATCH} = \frac{100(.95)}{1+8/60} = 83.8 \text{ moles/hr}$$

This results in a 11.73% increase in prod. rate. This increase is entirely dependent on reactor volume. As reactor volumes get bigger the downtime between batches becomes a smaller factor in production. For small reactors, batch may have a lesser prod. rate. The draw back is that batch reactors are more labor intensive than CSTR's.

5-8

Energy balance

$$\Delta\left(H + \frac{1}{2}\frac{V^2}{g_c} + \Delta Z\right) = Q - W$$

"the sum of internal, kinetic and potential energy change = heat added less work done

$$\int c_p dT = a\Delta T + \frac{b}{2}\left(T_2^2 - T_1^2\right) + \frac{c}{3}\left(T_2^3 - T_1^3\right)$$

1 lb of air

$$\Delta H = \frac{1}{29}\left[6.39(300) + \frac{9.8\times10^{-4}}{2}\left(760^2 - 460^2\right) - \frac{8.18\times10^{-8}}{3}\left(760^3 - 460^3\right)\right]$$

$$\Delta H = 71.966 \text{ Btu/lb}$$

$$\rho_1 v_1 = \rho_2 v_2 \qquad p = \rho RT$$

$$v_2 = v_1\left(\frac{P_1 T_2}{P_2 T_1}\right)$$

$$\frac{1}{2}\left(v_2^2 - v_1^2\right) = \frac{1}{2}v_1^2\left[\left(\frac{P_1 T_2}{P_2 T_1}\right)^2 - 1\right]$$

$$= \frac{1}{2}(100)^2\left[\left(\frac{(30)460}{(15)760}\right)^2 - 1\right]\frac{1}{32.2(778)}$$

$$= +.093 \text{ Btu/lb}$$

$$\frac{g\Delta Z}{g_c J} = -10\left(\frac{32.2}{32.2}\right)\frac{1}{778} = -.0129 \frac{\text{Btu}}{\text{lb}}$$

$$-Q = 71.966 - .093 - .0129 = 71.86 \frac{\text{Btu}}{\text{lb}}$$

$$Q_{TOTAL} = (71.86)(200) = 14372 \frac{\text{Btu}}{\text{hr}}$$

5-9

(a) for heat exchanger

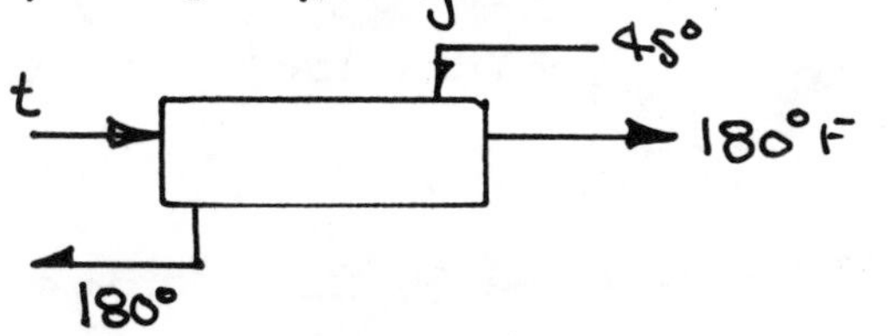

$Q = mc_p\Delta t = UA\Delta t_{lm}$

$\Delta t = t - 180$

$$\Delta t_{lm} = \frac{(t-180)-(180-45)}{\ln \frac{t-180}{135}} = \frac{(t-180)-135}{\ln \frac{t-180}{135}}$$

$mc_p\Delta t = UA\Delta t_{lm}$

has the form

$$K_1\phi = K_2 \frac{\phi - 135}{\ln \frac{\phi}{135}}$$

where $\phi = t - 180$

$$\frac{K_1}{K_2}\phi \ln \frac{\phi}{135} = \phi - 135$$

has solution $\phi = 135$ (by inspection)

$\therefore$ $t - 180 = 135$; $t = 315$

if feed is at 70°F H_{Rx} is used to heat entire mixture.

if m_o is original moles present, then for 40% conversion:

$$m_o(.4)H_{Rx} = m_o(.6)c_p\Delta t$$

there is a small error here due to unaccounted product moles, but since the product has such a high molecular weight product moles can be neglected precisely

$$m_o(.4)H_{Rx} = \left[m_o(.6) + \frac{.4m_o(104)}{140000}\right]c_p\Delta t$$

solving the simpler equation

5-9 (con't)

$$H_{Rx} = \frac{.6}{.4}c_p\Delta t = \frac{.6}{.4}(.45)(315-70)$$

$H_{Rx} = 165.4$ cal/gmole

(b) $$\tau = \frac{C_{A_0}X_A}{-r_A} = \frac{.9423 \frac{\text{moles}}{\text{liter}}(.4)}{.869 \text{ moles/liter-sec}}$$

$\tau = 0.4337$ sec

$$v_o = 941 \frac{g}{sec} \Big/ \left[.9423 \frac{\text{moles}}{\text{liter}} \times 104 \frac{g}{\text{mole}}\right]$$

$v_o = 9.602$ liter/sec

$V = v_o\tau = 9.602(.4337) = 4.16$ liters

5-10 basis

1 mole C_4H_{10};

$$C_4H_{10} + \frac{13}{2}O_2 \rightarrow 4CO_2 + 5H_2O$$

at 100% excess with x moles CO_2 formed

$$C_4H_{10} + 13O_2 + 13\left(\frac{79}{21}\right)N_2 \rightarrow$$

$$xCO_2 + (4-x)CO + 5H_2O + \frac{17-x}{2}O_2 + 13\left(\frac{79}{21}\right)N_2$$

$$\Delta H_{Rx} = xH_{f_{CO_2}} + (4-x)H_{f_{CO}} + 5H_{f_{H_2O}} - H_{f_{C_4H_{10}}}$$

$$\Delta H_{Rx} = x(-94.0518) + (4-x)(-26.4157) + 5(-57.798) - (-29.812)$$

$\Delta H_{Rx} = -67.6361x - 364.8408$ kcal

heat of reaction used to heat products and excess reactants

$$Q = \sum m \int_{T_1}^{T_2} c_p dT = \sum m_i\left[a_i(T_2 - T_1) + \frac{b_i}{2}(T_2^2 - T_1^2) + \frac{c_i}{3}(T_2^3 - T_1^3)\right]$$

5-10 (con't)

$$\Delta Q = (1124-298)[x(6.3930)+(4-x)(6.324) + 5(7.219) + \frac{17-x}{2}(6.0954) + 48.905(6.442)]$$
$$+ \frac{10^{-3}}{2}(1124^2-298^2)[x(10.1)+(4-x)(1.863) + 5(2.374) + \frac{17-x}{2}(3.2533) + 48.905(1.4125)]$$
$$+ \frac{10^{-6}}{3}(1124^3-298^3)[x(-3.405)+(4-x)(-.2801) + 5(0.267) + \frac{17-x}{2}(-1.0171) + 48.905(-.0807)]$$

$\Delta Q = .2064x + 416139.7$ cal

$= .2064x + 416.1397$ kcal

$\Delta H_{Rx} + \Delta Q = 0$

$.2064x + 416.1397 = 67.6361x + 364.8408$

$x = 0.7608$

or 76.08% C_4H_{10} burned to CO_2

5-11

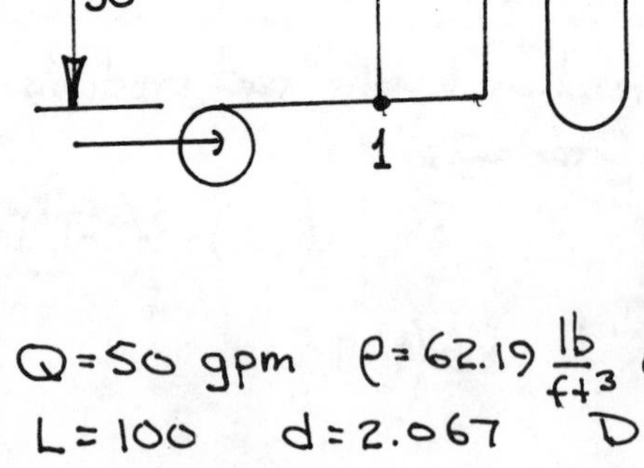

(a) $Q = 50$ gpm $\quad \rho = 62.19\ \frac{lb}{ft^3}$ @ 80°F

$L = 100 \quad d = 2.067 \quad D = .172$ ft

$\varepsilon = 0.0002$

$N_{Re} = 50.6\ \frac{Q\rho}{d\mu}$

5-11 (con't)

$N_{Re} = 88512$

Sacham equation for f yields

$f = .02297$

$v_1 = v_2 = 0.408\ \frac{Q}{d^2} = 4.1775\ \frac{ft}{sec}$

Bernoulli equation for 1 → 2

$$\Delta Z + \frac{\Delta v^2}{2g} + \frac{\Delta P}{\rho} \cdot 144 + h_{f12} = 0$$

$\Delta Z = 30 \quad \Delta v^2 = 0 \quad P_2 = 20$

$$\therefore\ 30 + \frac{20-P_1}{\rho}(144) + h_{f12} = 0$$

$$h_{f12} = f\left(\frac{L}{D}\right)\frac{v_2^2}{2g} = 4.728$$

$$P_1 = (30+4.728)\frac{\rho}{144} + 20$$

$P_1 = 34.998$ psig

$H_{41} = P_1(144)/\rho = 81.038$ ft

$L_4 = H_{41} - 30 = 51.038$ ft

(b) to find time until flow drops to 95% requires recalculating P_1 at new flow

$Q = 0.95(50) = 47.5$ gpm

$v_2 = v_1 = .408\ \frac{Q}{d^2} = 4.536$ ft/sec

$v_s = .408\ \frac{Q}{48^2} = .00841$ ft/sec @ 47.5 gpm

assume f remains same

Check:

Sacham equation yields

$f = 0.02309$ (use new f)

5-11 (con't)

$h_{f12} = f\left(\frac{L}{D}\right)\frac{v_2^2}{2g} = 4.289$

Bernoulli equation yields

$P_1' = (30 + 4.289)\frac{\rho}{144} + 20$

$P_1' = 34.808$ psig

$H_{S1} = P_1' \, 144/\rho = 80.598$

$L_S = H_{S1} - 30 = 50.598$ ft

Level drop: $L_4 - L_5 = 0.44$ ft

$v_4 = .408 \frac{50}{48^2} = .00885 \frac{ft}{sec}$

$v_{average} = \frac{v_4 + v_5}{2} = 0.00863 \frac{ft}{sec}$

$t = \frac{\text{Level drop}}{v_{average}} = \frac{0.44}{.00863}$

$t = 51$ sec

5-12

$\ln K = \frac{5038.8}{698} - .09415 \ln 698 + 0.0014553$
$(698) - 2.5(10^{-6})\, 698^2 - 5.269$

$K = 3.0993$

$K = \frac{[CO_2]\,[H_2]}{[CO]\,[H_2O]}$

let X = moles CO react
= moles CO_2 produced
= moles H_2 produced

$K = \frac{x\,(.7 + x)}{(.3 - x)(.6 - x)} = 3.0993$

5-12 (con't)

solving the quadratic

$X^2 - 1.6622X + .2657 = 0$

$X = 0.1792$

= moles H_2 produced/mole CO

5-13

For immiscible solvents
use Kremser-Brown relationship

$\varepsilon = \frac{10(150)}{100} = 15$

$E = \frac{\varepsilon^{n+1} - \varepsilon}{\varepsilon^{n+1} - 1}$; $E = 1 - 1/(5)(238)(1000)$
$= 0.9999915966$

$n = \left[\ln\left((\varepsilon - E)/(\varepsilon - \varepsilon E)\right)\right] / \ln \varepsilon$

$n = \frac{\ln\left[\frac{15 - .9999\cdots}{15 - 15(.9999\cdots)}\right]}{\ln 15} = 4.29$ stages

use 5 stages

5-14

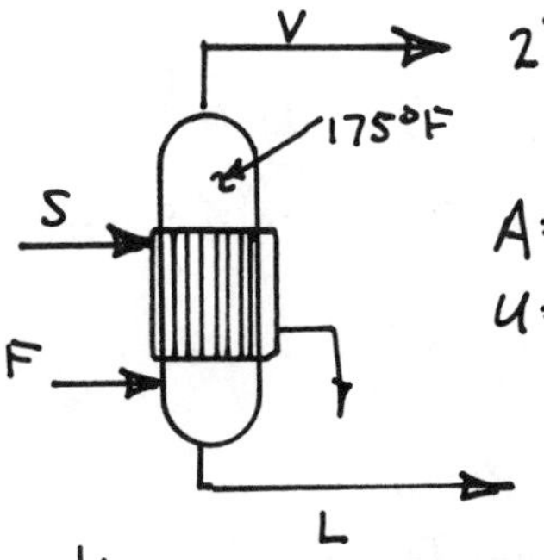

$2''$ Hg $\cong$ 1 psia

$A = 460\ ft^2$

$U = 300 \frac{BTU}{hr\text{-}ft^2\text{-}°F}$

This problem is an engineering judgement problem. Judgement is whether BPE data or heat capacity data are valid since problem can be solved with either.

Assume:
1. no heat loss
2. no subcooling of steam
3. no fouling
4. Δt constant along length

5-14 (con't)

heat bal.:

$t_{sat} = 350.0°F$ @ 134.7 psia

$Q = UA\Delta t = 300(460)(350-175)$

$Q = 2.415 \times 10^7$ Btu/hr

mat bal.:

$F = V + L$

let x = fract of feed converted to vapor

w = weight% conc. NaOH

$V = xF$

NaOH

$20F = wL$

$(1-x)F = L$

$\therefore$ $w = 20/(1-x)$

Assume BPE data correct.

$t_{sat} = 101.7°F$ @ 1 psia

$BPE = 175 - 101.7 = 73.3°F$

from data:

$BPE = 1.5(wgt\%) - 20$

$\therefore$ $wgt\% = (BPE+20)/1.5$; $wgt\% = 62.2\%$

$w = 20/(1-x) = 62.2$

$x = .67846$

water evaporated:

$40000(.67846) = 27138$ lb/hr

Assume specific heat data correct.

$Sh_g + Fh_{feed} = Sh_{cond} + Fxh_v + F(1-x)h_L$

$S(h_g - h_{cond}) = S\lambda = 24150000$

$= F(xh_v + (1-x)h_L - h_{feed})$

superheat tables at 1 psia 175°F

$h_v = 1138$ Btu/lb

5-14 (con't)

$h_{feed} = (1 - .032\sqrt{20})110 = 94.26 \frac{Btu}{lb}$

$h_{Liquor} = (1 - .032\sqrt{20/(1-x)})(175)$

$= [1 - .1431\sqrt{1-x}/(1-x)]175$

substituting into heat bal. (& simplifying)

$\Delta = 0 = 5.50771x - .14131\sqrt{1-x} - 2.99059$

trial & error

$x \cong .4751$

water evap.

$xF = .4751(40000) = 19004 \frac{lb}{hr}$

$wgt\% = w = \frac{20}{1-x} = 38.1\%$

BPE data are suspect since straight BPE data occur only in dilute region. Specific heat data are probably more reliable.

5-15

The extent of this problem is so vast that the student should never consider solving a problem such as this on the exam. The solution is presented here as information only.

5-15 (con't)

5 psig

W	ΔW	Δt	Δt/ΔW	$\bar{W}$
0	-	-	-	-
2	2	24	12	1
4	2	47	23.5	3
6	2	75	37.5	5
8	2	98	49	7
10	2	128	64	9
12	2	152	76	11
14	2	166	83	13
16	2	198	99	15
18	2	300	150	17

15/30/50 psig

W	ΔW	$\bar{W}$
0	-	-
5	5	2.5
10	5	7.5
15	5	12.5
20	5	17.5
25	5	22.5
30	5	27.5
35	5	32.5

$\bar{W}$	15 psig		30 psig		50 psig	
	Δt	Δt/ΔW	Δt	Δt/W	Δt	Δt/ΔW
-	-	-	-	-	-	-
2.5	50	10	26	5.2	19	3.8
7.5	131	26.2	72	14.4	49	9.8
12.5	204	40.8	113	22.6	74	14.8
17.5	275	55	150	30.0	99	19.8
22.5	349	69.8	194	38.8	127	25.4
27.5	434	86.8	233	46.6	156	31.2
32.5	674	134.8	295	59.0	178	35.6

5-15 (con't)

PLOT Δt/ΔW vs $\bar{W}$

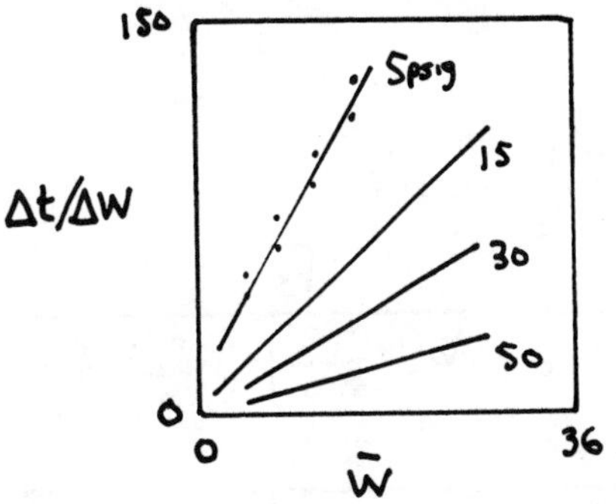

$$\frac{\Delta t}{\Delta W} = K_p V + B \; ; \; \rho V = W ; \; \therefore \frac{\Delta t}{\Delta W} = K_p \frac{W}{\rho^2} + \frac{B}{\rho}$$

$$\therefore \frac{\Delta t}{\Delta W} = K_p' W + B'$$

Linear regression yields data below

ΔP	K_p'	B'	comment
5	6.173	6.119	last point not used
15	3.023	2.757	"
30	1.643	1.615	"
50	1.063	1.457	all points used

Definitions:

C = mass of particles deposited on filter/vol. filt.

C_S = mass of solids in slurry/volume fed to filter

V_c = volume of cake

V_p = volume of particles in cake

V_{FC} = volume of filtrate in cake

V_F = volume of filtrate

ρ = density of filtrate = 62.4 lb/ft^3

f_s = mass fraction solids in slurry

m_f/m_c = mass ratio: wet cake to dry cake

ρ_c = density of cake

ε = void fraction in cake

BASIS: 1 lb of slurry

weight of dry cake: f_s

$V_c = f_s/\rho_c$

weight filtrate in cake = $f_s(m_f/m_c)$

$V_{FC} = (m_f/m_c) f_s/\rho$

$$\varepsilon = \left[(m_f/m_c) f_s/\rho\right] = \frac{m_f}{m_c} \frac{\rho_c}{\rho}$$

5-15 (con't)

$V_p = V_c(1-\varepsilon)$

$V_F = (1 - f_s \frac{m_f}{m_c} - f_s)/\rho$

$$C_S = \frac{f_s}{V_p + V_F + V_{Fc}} = \frac{f_s}{V_c\left(1-\frac{m_f \rho_c}{m_c \rho}\right) + \left(1-\frac{m_f f_s}{m_c} - f_s\right)\frac{1}{\rho} + \frac{m_f f_s}{m_c \rho}}$$

$$= \frac{1}{\frac{1}{\rho_c}\left(1-\frac{m_f \rho_c}{m_c \rho}\right) + \left(\frac{1}{f_s} - \frac{m_f}{m_c} - 1\right)\frac{1}{\rho} + \frac{m_f}{m_c}\frac{1}{\rho}}$$

$$C = \frac{C_S}{1 - \left(\frac{m_f}{m_c} - 1\right)\frac{C_S}{\rho}}$$

ΔP	C_S	C
5	11.17	12.49
15	11.18	12.21
30	11.18	12.21
50	11.20	12.23

$$K_p' = \frac{C \alpha \mu}{A^2 (\Delta P) g \rho^2} \qquad B' = \frac{R_m \mu}{A \Delta P g \rho}$$

using regression values for K_p' & B'

ΔP	α ft/lb	R_m ft^{-1}
5	5.95×10^{12}	1.317×10^{10}
15	8.75×10^{12}	1.78×10^{10}
30	9.51×10^{12}	2.086×10^{10}
50	10.26×10^{12}	2.823×10^{10}

linear regression of formula

$$\ln \alpha = S \ln \Delta P + \ln \alpha_o$$

yields $\alpha_o = 1.319 \times 10^{12}$

$S = 0.2357$

$\therefore \alpha = 1.319 \times 10^{12} (\Delta P)^{.2357}$

$\Delta P [\equiv \text{lb/ft}^2]$

5-15 (con't)

linear regression of

$$R_m = t \, \Delta P + R_{m_o}$$

yields

$R_{m_o} = 1.2 \times 10^{10}$

$t = 2.226 \times 10^{6}$

$\therefore R_m = 2.226 \times 10^6 \Delta P + 1.2 \times 10^{10}$

LARGER FILTER:

$V_c = 10(1.5/12) = 1.25$ ft^3

$W_c = 1.25(73) = 91.25$ lb (dry)

$W_c = 91.25(1.47) = 134.14$ lb (wet)

$W_{FC} = 134.14 - 91.25 = 42.8875$ lb

$V_{FC} = 42.8875/62.4 = .687$ ft^3

$\varepsilon = .687/1.25 = 0.5498$

$V_p = 1.25(1 - .5498) = .562$ ft^3

$V_{\text{TOTAL SLURRY}} = 91.25/11.18 = 8.162$ ft^3

$V_F = 8.162 - .562 - .688 = 6.912$ ft^3

$W_F = 6.912(62.4) = 431.3$ lb

$$K_p = \frac{C \alpha \mu}{A^2 \Delta P g \rho^2}$$

$$K_p' = \frac{12.21(1.319 \times 10^{12})(25 \times 144)^{.2357} \, 6.72 \times 10^{-4}}{10^2 (25 \times 144)(32.2)(62.4)^2}$$

$= 1.652$

$$B' = \frac{R_m \mu}{A \Delta P g \rho}$$

$$B' = \frac{[2.226 \times 10^6 (25 \times 144) + 1.2 \times 10^{10}] \, 6.72 \times 10^{-4}}{10 (25 \times 144)(32.2)(62.4)}$$

$B' = .18593$

S-15 (cont)

$$t = \int_0^W (K_p' W + B')\,dW$$

$$= 1.652\,\frac{W^2}{2} + .18593\,W$$

for $W = 431$

$t = 2562$ min $= 42.7$ hours

one cycle filtrate $= 431.3$ lbs

$$\left(\frac{dt}{dW}\right)_{W=431.3} = 1.652(431.3) + .18593$$

$$= 712.5 \text{ sec/lb}$$

$$\left(\frac{dW}{dt}\right)_{WASH} = 0.8\,\frac{1}{712.5} = 0.0011228\ \frac{\text{lb}}{\text{sec}}$$

$t_{WASH} = 431.3/.0011228$

$= 384128$ sec

$= 6402$ min

$= 106$ hours (WHEW!)

S-16

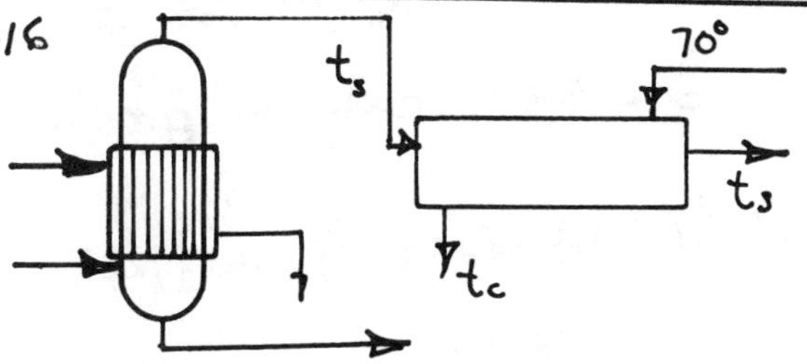

$Q_{cooling\ water} = 48000\,c_p(t_c - 70)$

$\Delta h_{vapor} = 24000\,\lambda_{sat}$

$\Delta Q = Q_{cooling\ water} = UA\,\Delta t_{lm}$

$$= \frac{(t_s - 70) - (t_s - t_c)}{\ln \frac{t_s - 70}{t_s - t_c}}\,UA$$

$$= UA\,\frac{t_c - 70}{\ln \frac{t_s - 70}{t_s - t_c}}$$

S-16 (con't)

$$\therefore\ 48000(1)(t_c - 70) = 500(665.4)\,\frac{t_c - 70}{\ln \frac{t_s - 70}{t_s - t_c}}$$

$$\ln \frac{t_s - 70}{t_s - t_c} = 6.931$$

$$\frac{t_s - 70}{t_s - t_c} = 1024$$

$$t_s - t_c = \frac{t_s - 70}{1024}$$

$$\Delta Q = 2400\,\lambda_s = 500(665.4)\,\frac{t_s - 70 - \frac{t_s - 70}{1024}}{6.931}$$

$$\lambda_{sat} = 1.998\,t_s - 139.56$$

find λ_{sat} & t_{sat} from steam tables that satisfy equation

t_s	λ_s table	λ_s equation
100	1037	60.24
200	977.9	260.04
300	910.1	459.84
400	826.0	659.64
450	774.5	759.54
455	768.9	769.53

$t_s = 455$ °F

$p = 444$ psia

5-17

$X_A = 0$: $C_A = 50$ vols, $C_i = 50$, $C_R = 0$

$X_A = 1$: $C_A = 0$ vols, $C_i = 50$, $C_R = 150$

$$\varepsilon_A = \frac{200-100}{100} = 1$$

$$\tau = \frac{1}{C_{A_0}K}\left[2\varepsilon_A(1+\varepsilon_A)\ln(1-X_A) + \varepsilon_A^2 X_A + \frac{X_A(\varepsilon+1)^2}{1-X_A}\right]$$

for $\varepsilon_A = 1$

$$\tau = \frac{1}{C_{A_0}K}\left[4\ln(1-X_A) + X_A + \frac{4X_A}{1-X_A}\right]$$

$$\tau = \frac{1}{.0625(0.01)}\left[4\ln(1-.9) + .9 + \frac{4(.9)}{1-.9}\right] = 44303 \text{ sec}$$

$$v_0\tau = V = 1(44303) = 44303 \text{ liters}$$

5-18

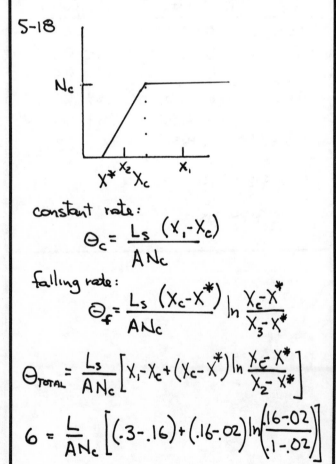

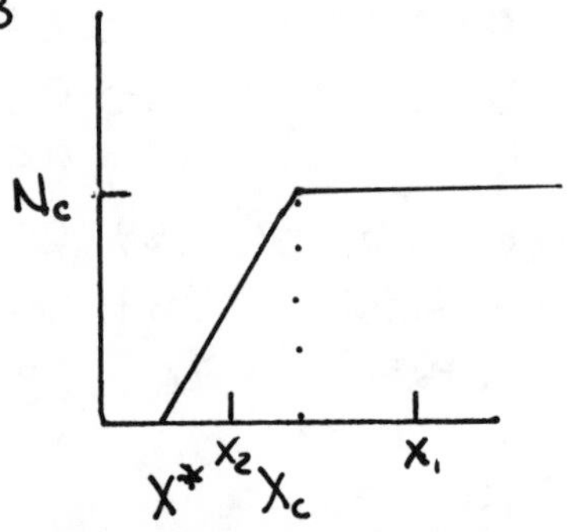

constant rate:

$$\Theta_c = \frac{L_s(X_1 - X_c)}{AN_c}$$

falling rate:

$$\Theta_f = \frac{L_s(X_c - X^*)}{AN_c}\ln\frac{X_c - X^*}{X_3 - X^*}$$

$$\Theta_{TOTAL} = \frac{L_s}{AN_c}\left[X_1 - X_c + (X_c - X^*)\ln\frac{X_c - X^*}{X_2 - X^*}\right]$$

$$6 = \frac{L}{AN_c}\left[(.3-.16) + (.16-.02)\ln\left(\frac{.16-.02}{.1-.02}\right)\right]$$

5-18 (con't)

$$\frac{L}{AN_c} = 27.479$$

$$\Theta_{NEW} = 27.479\left[(.35-.16) + (.16-.02)\ln\frac{.16-.02}{.06-.02}\right]$$

$$\Theta_{NEW} = 10.04 \text{ hrs}$$

5-19

air as an ideal gas

$C_p = 7$ cal/gmole°K

$$n = \frac{PV}{RT} = \frac{(1\text{atm})(30000\ \frac{\ell}{\text{min}})}{.08205(298°\text{K})} = 1227\ \frac{\text{gmole}}{\text{min}}$$

$n\Delta H = Q - W \qquad \Delta S = 0$

$$c_p \ln\frac{T_2}{T_1} + \frac{\Delta Q}{T_1} = 0$$

$$W = T\Delta S - \Delta H$$

$$\Delta H = C_p(T_2 - T_1) = 7(15-38) = -161\ \frac{\text{cal}}{\text{gmole}}$$

$$\Delta S = C_p \ln T_2/T_1 = (7)\ln\frac{273+15}{273+38}$$

$$= -.5378 \text{ cal/°K-gmole}$$

$$W = [(273+38)(-.5378) + 161]\,1227$$

$$= -7686 \text{ cal/min}$$

$$\text{Power} = 7686\ \frac{\text{cal}}{\text{min}} = 536 \text{ watts}$$

5-20 (all dollars in thousands)

Use PW Method

$$PW_{BATCH} = (96-20)(P/A,15\%,5)(1-.48)$$

(96 = PROFIT (REVENUE); 20 = EXPENSE)

$$+\ 2(P/F,15\%,5) - 27$$

(2 = salvage; 27 = INVESTMENT)

$$+\ \frac{27-2}{5}(P/A,15\%,5)(.48)$$

($\frac{27-2}{5}$ = DEP. CREDIT)

$$= \$114.5^k$$

$$PW_{CONTINUOUS} = (99.5-19)(P/A,15\%,5)(1-.48)$$

$$+\ 1.9(P/F,15\%,5) - 55$$

$$+\ \frac{55-1.9}{5}(P/A,15\%,5)(.48)$$

$$= \$103.36k$$

Batch better: PW is greater

6-1

Bernoulli:

(a) $\Delta h_z + h_{pump} + h_f + \Delta h_p^{\,0} = 0$ (with Δh_p struck out)

$65 + h_{pump} + h_f = 0$

$h_f = f\left(\frac{L}{D}\right) 0.00259 \frac{Q^2}{d^4}$

Q ≡ gpm d ≡ inches D ≡ feet

d = 2.067

$\therefore \Delta = 0 = 65 + h_{pump} + 0.18122 f Q^2$

$N_{Re} = 50.6 \frac{Q\rho}{d\mu}$

$\rho = 62.4 \quad \mu = 1.43\ cp$

$N_{Re} = 1068.2\ Q \qquad \varepsilon/D = 0.0008$

Sacham equation

$$f = \left\{-2\log\left[\frac{\varepsilon/D}{3.7} - \frac{5.02}{N_{Re}}\log\left(\frac{\varepsilon/D}{3.7} + \frac{14.5}{N_{Re}}\right)\right]\right\}^{-2}$$

becomes

$$f = \left\{-2\log\left[2.162\times10^{-4} - \frac{.0047}{Q}\log\left(2.162\times10^{-4} + \frac{.1357}{Q}\right)\right]\right\}^{-2}$$

Q	f	Δ
96	.02144	-18.39
97	.02142	-15.88
98	.0214	-13.36
99	.02137	-10.84
100	.02135	-8.31
101	.02133	-5.37
102	.02131	-2.43
103	.02129	0.52 ←
104	.02126	3.48
105	.02124	6.45
106	.02122	9.42

flow = 103 gpm

$\varepsilon_p = .48 \quad \varepsilon_D = 0.8$

(b) driver hp $= \frac{Qh\rho}{247000\,\varepsilon_p \varepsilon_D} = 7.142$ hp

6-2

This is a problem involving differential equations

heat balance: use 80°F as ref.

In + Out = Accum

$100(.94)(80-80) - 80(.94)(t-80) = \frac{d}{d\theta}\left(m(.94)(t-80)\right)$

m = lb in tank at time θ

$80(80-t) = \frac{d}{d\theta}\left[m(t-80)\right]$

$6400 - 80t = m\frac{dt}{d\theta} + (t-80)\frac{dm}{d\theta}$

material balance:

$100 - 80 = \frac{dm}{d\theta}$

$\frac{dm}{d\theta} = 20; \quad m - 400 = 20\theta$

Substitute in heat bal. & simplify

$400 - 5t = (\theta + 20)\frac{dt}{d\theta}$

$$\int_0^\theta \frac{d\theta}{\theta+20} = \int_{400}^{t} \frac{dt}{400-5t}$$

$$\ln(\theta+20)\Big|_0^\theta = -\frac{1}{5}\ln(400-5t)\Big|_{400}^{t}$$

$$\left(\frac{\theta+20}{20}\right)^5 = \frac{1600}{5t-400}$$

$$t = \frac{320}{(.05\theta+1)^5} + 80$$

θ = 6

$$t = \frac{320}{[(.05)6+1]^5} + 80 = 166.19°F$$

6-3

(a) $-r_A = kC_AC_P$

$C_o = \text{constant} = C_A + C_P$

$-r_A = kC_A(C_o - C_A) = k(C_AC_o - C_A^2)$

$\frac{dr_A}{dC_A} = 0 = k(C_o - 2C_A)$

C_A at max rate $= kC_o/2 = 0.5$

$r_{a\,max} = -k(C_AC_o - C_A^2)$

$= -(1)(.5(1) - .5^2) = .025 \frac{\text{mole}}{\text{liter-min}}$

(b) integrated form: $\ln \frac{C_P/C_{Po}}{C_A/C_{Ao}} = C_okt$

at $C_P = 0.9$; $C_A = 1$

$\ln \frac{.9/.01}{.1/.99} = 6.792 = (1)(1)t$

$t = 6.792$ min

(c) at max rate

$C_A = C_P = .5$

$\ln \frac{0.5/0.01}{0.5/0.99} = (1)(1)t$

$t = 4.595$ min

6-4

(a)

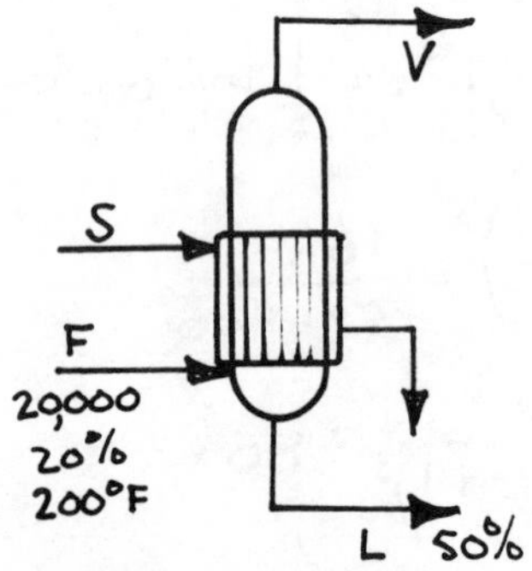

steam table

STATE	PRESS	TEMP	h_f	h_{fg}	h_g
Sat	2	126.08	93.99	1027.2	1116.2
Sat	134.63	350	321.63	870.7	1192.3
Superh	2	176.08	—	—	1139.6
	psia	°F	Btu/lb	Btu/lb	Btu/lb

6-4 (con't)

NaOH $C_p = 1 - .032\sqrt{\%}$

C_p 20%: .8569

C_p 50%: .7737

mass bal:

$20000 = L + V$

$20000(.2) = (.5)L$

$V = 12000$

$L = 8000$

heat bal:

$20000(.8569)(200-32) + S(870.7)$

$= 12000(1139.6) + 8000(.7737) \times (176.8-32)$

$S = 13428.6$ lb/hr

$Q = 13675200 + 896254$

$= 14571454$ Btu/hr

$Q = UA\Delta t = 1457145 = 300A(350-176)$

$A = 279.3\ ft^2$

(b) steam economy $= \frac{\text{lb vaporized}}{\text{lb steam}}$

$= 12000/13428.6 = 0.8936$

6-5 BASIS 1 mole mixture

stoichiometry 30% excess

$0.9CH_4 + .06C_2H_6 + [.04 + 1.3(7.56)]N_2 + 2.01(1.3)O_2 \rightarrow$

$1.02CO_2 + 1.98H_2O + [.04 + (1.3)7.56]N_2 + .3(2.01)O_2$

	lbs		lbs
CH_4	14.4387	CO_2	44.89
C_2H_6	1.804	H_2O	35.67
N_2	276.436	O_2	19.295
O_2	83.613	N_2	276.436

LHV $= 21520(14.4387) + 20432(1.804)$

$= 347580$ Btu/mole mixture

th. flame temp $= \text{LHV}/\Sigma Wc_p + 90$

6-5 (con't)

t guess	C_p CO_2	H_2O	O_2	N_2	$\sum WC_p$	calc. flame temp.
4000	.302	.582	.27	.29	120.5	2974
3000	.290	.550	.26	.28	115.5	3099
3100	.290	.552	.262	.281	115.8	3090

th. flame temp ≅ 3095 °F

6-6

$h1_g = 1306.9$
$S1_g = 1.5894$ } superheated 400psi 600°F

$h2_f = 218.82 \quad h2_{fg} = 945.3$

$h2$ = exit enthalpy

$S2_f = 0.3680 \quad S2_{fg} = 1.3313$

$S2$ = exit entropy

$S2 = S1_g = 1.5894 = S2_f + X\,S2_{fg}$

$X = \dfrac{S1_g - S2_f}{S2_{fg}}$

$X = .91745$

$h2 = h2_f + h2_{fg}X$

$h2 = 1086$ Btu/lb

work $= h1_g - h2 = 220.82 \ \frac{\text{Btu}}{\text{lb}}$

steam rate $= 2544.5/220.82$
$= 11.523$ lb steam/hp-hr

actual rate $= \dfrac{11.523}{.75} = 15.364 \ \dfrac{\text{lb steam}}{\text{hp-hr}}$

6-7

1st order PFR

$K\tau = KC_{A_0}V/F_{A0} = -(1+\varepsilon)\ln(1-X_A) - \varepsilon_A X_A$

$F_{A_0} = 4$ lbmoles/hr

$C_{A_0} = P_{A_0}/RT = 4.6/(.7302(1660))$
$= .0038$ lbmole/ft^3

$\varepsilon_A = (7-4)/4 = .75$

$V = \dfrac{F_{A0}}{KC_{A_0}}\left[-(1+\varepsilon_A)\ln(1-X_A) - \varepsilon_A X_A\right]$

$V = \dfrac{4}{10(.0038)}\left[-(1+.75)\ln(1-.8) - .75(.8)\right]$

$V = 233.6$ ft^3

6-8

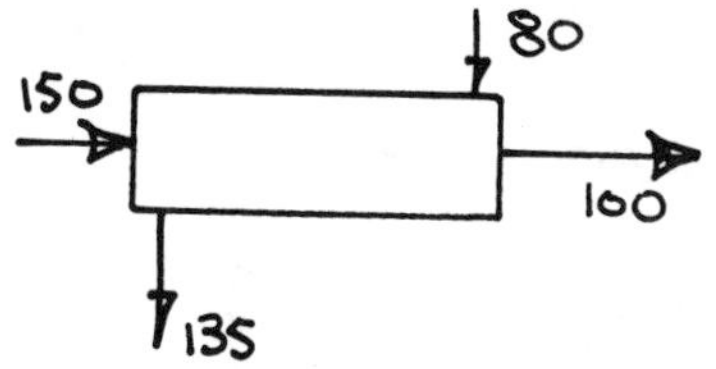

$Q = 15000(1)(135-80) = 825000$

$\Delta t_{lm} = \dfrac{(100-80)-(150-135)}{\ln\frac{20}{15}} = 17.38$ °F

$U_o = \dfrac{1}{\frac{1}{h_o} + \frac{D_o}{D_i h_{di}} + \frac{1}{h_{do}} + \frac{D_o}{D_i h_i} + \frac{X_{wall} D_o}{K_m \bar{D}_L}}$

$D_o/D_i = 1.315/1.049 = 1.2536$

$\bar{D}_L = \dfrac{D_o - D_i}{\ln D_o/D_i} = \dfrac{1.315-1.049}{\ln 1.315/1.049} = 1.1770$ inches

$X_{wall} = .133/12 = .0111$ ft

$D_o/\bar{D}_L = 1.315/1.177 = 1.1172$

$U_o = \dfrac{1}{\frac{1}{300} + \frac{1.2536}{1000} + \frac{1}{500} + \frac{1.2536}{180} + \frac{(.0111)1.1172}{26}}$

$U_o = 71.28$

$Q = U_o A_o \Delta t_{lm} = 71.28 A_o (17.38) = 825000$
$A_o = 665.9$ ft^2

6-9

Since equil line is straight solve analytically

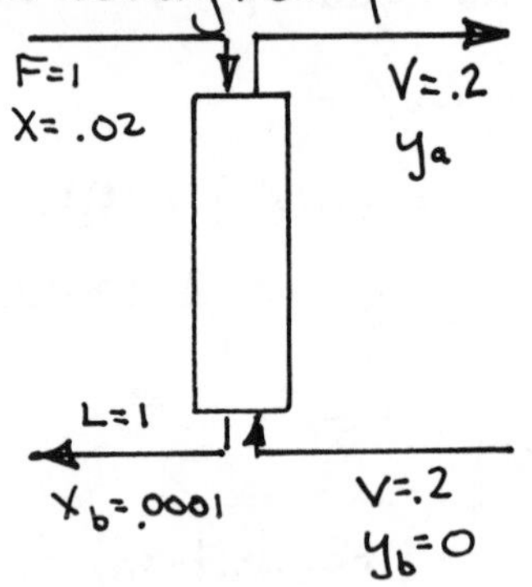

$F \cong L = 1$

$V = 0.2$

$V(y_a - y_b) = L(x_a - x_b)$

$.2(y_a - 0) = 1(.02 - .0001)$

$y_a = 0.0995$

$y_a^* = 9(.02) = .18$

$y_b^* = 9(.0001) = .0009$

$$N = \frac{\ln \dfrac{y_b - y_b^*}{y_a - y_a^*}}{\ln \dfrac{y_b - y_a}{y_b^* - y_a^*}}$$

N = 7.6 ideal plates

6-10

to obtain preliminary result for ΔP_m, a plot of ΔP vs t extrapolated to 0=t gives $\Delta P_m \cong 3.8$ see plot "PROB6-10a". A log-log plot of $\Delta P - \Delta P_m$ vs t gives a satisfactory straight line See plot "PROB6-10b".

$$R_m = \frac{-\Delta P_m g_c}{\mu u} = \frac{(3.8)144(32.2)}{.92(6.72\times10^{-4})(.0016)} = 1.78\times10^{10} \text{ ft}^{-1}$$

6-10 (con't)

from plot "PROB6-10b" regression analysis indicates that

$(1 - S) = 0.713349$

$S = 0.286651$

$(15.704)(\Delta P - \Delta P_m)^{.713349} = t$

$-(\Delta P - \Delta P_m)^{1-S} = K_r t$

$K_r = \frac{1}{15.704} = .0637$

$K_r = \alpha_o \mu u^2 c / g_c$

$$\alpha_o = \frac{K_r g_c}{\mu u^2 c} = \frac{.0637(32.17)}{.92\times6.72\times10^{-4}(.0016)^2(1.08)}$$

$\alpha_o = 1.199 \times 10^9$ ft/lb

6-11

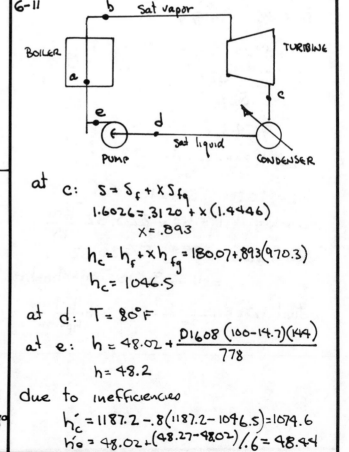

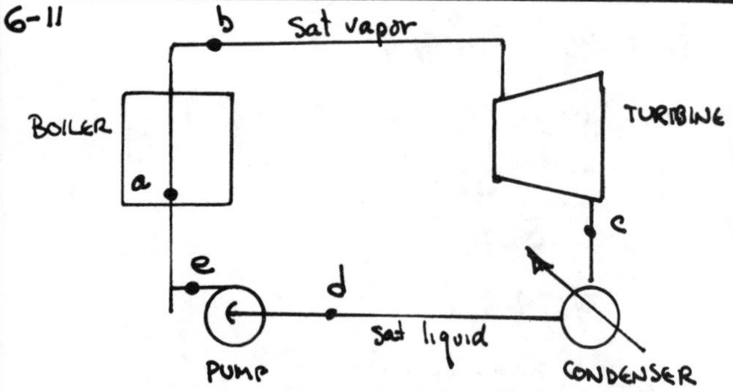

at c: $s = s_f + x s_{fg}$

$1.6026 = .3120 + x(1.4446)$

$x = .893$

$h_c = h_f + x h_{fg} = 180.07 + .893(970.3)$

$h_c = 1046.5$

at d: T = 80°F

at e: $h = 48.02 + \frac{.01608(100-14.7)(144)}{778}$

$h = 48.2$

due to inefficiencies

$h_c' = 1187.2 - .8(1187.2 - 1046.5) = 1074.6$

$h_e' = 48.02 + (48.27 - 48.02)/.6 = 48.44$

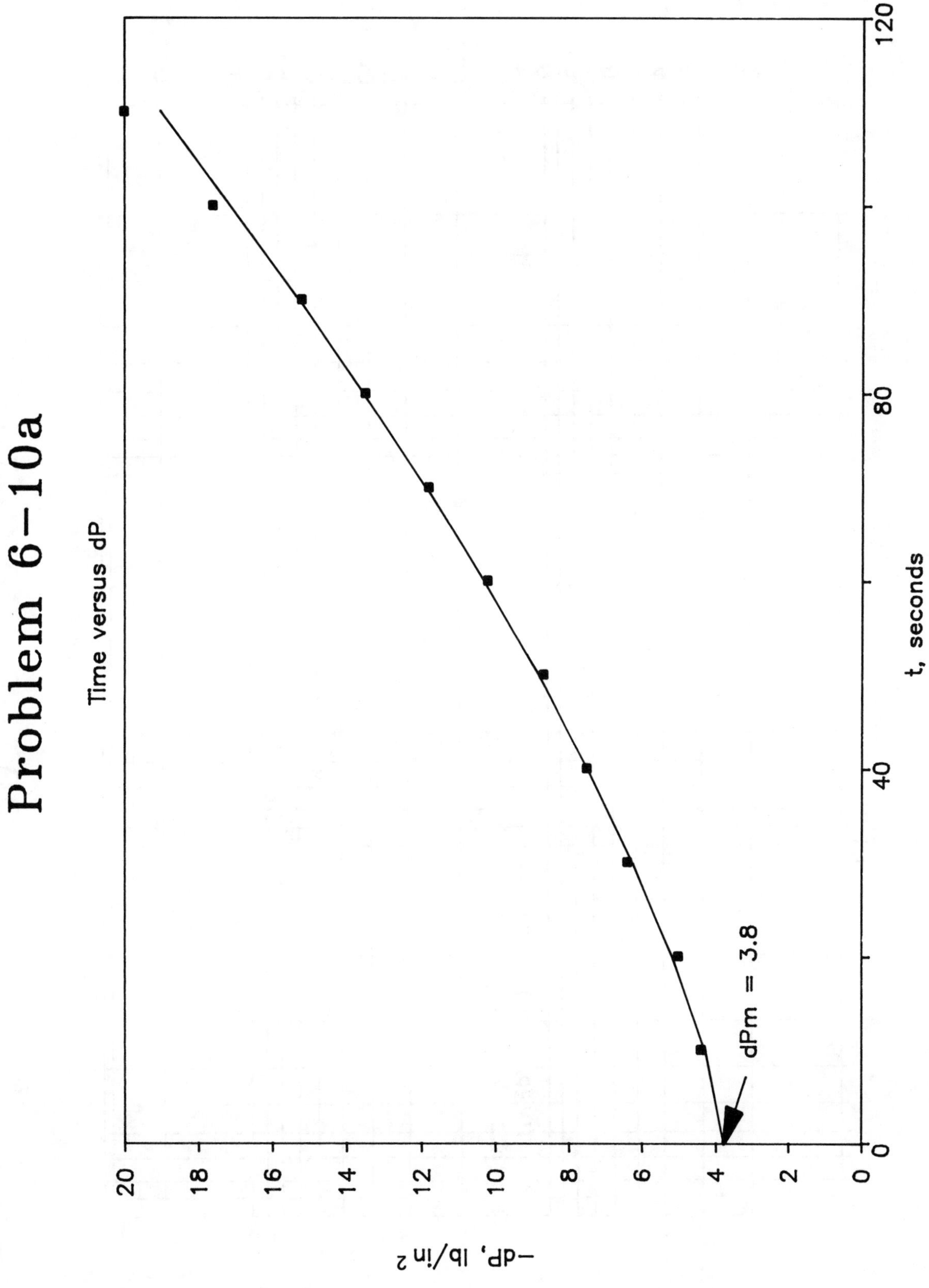
Problem 6–10a
Time versus dP
dPm = 3.8
t, seconds
−dP, lb/in²
0
40
80
120
0
2
4
6
8
10
12
14
16
18
20

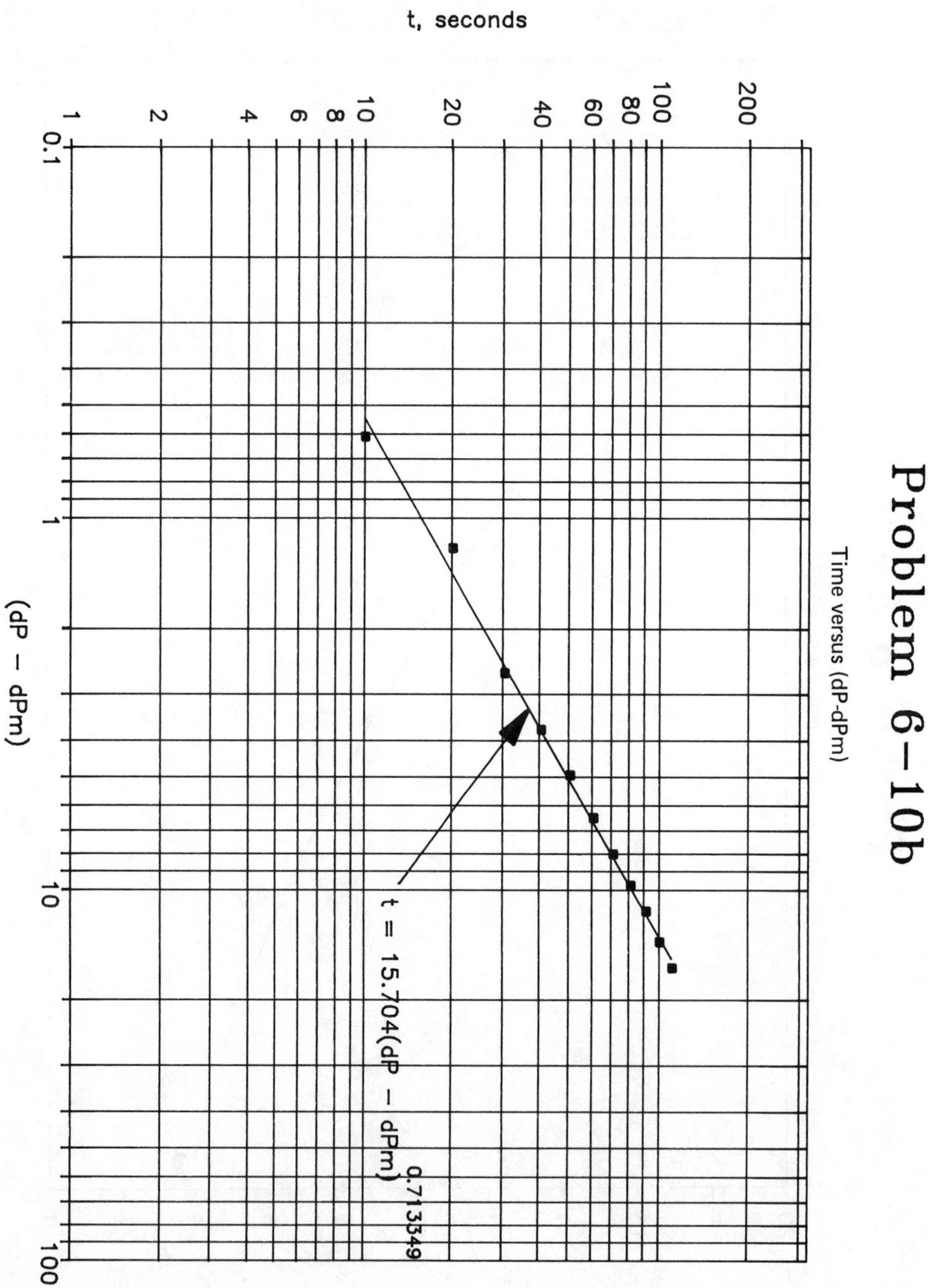
Problem 6-10b
Time versus (dP-dPm)
t, seconds
1
2
4
6
8
10
20
40
60
80
100
200
0.1
1
10
100
(dP - dPm)
t = 15.704(dP - dPm)^0.713349

6-11 (con't)

$$\eta_{th} = \frac{(1187.2-1074.6)-(48.44-48.02)}{(1187.2-48.44)}$$

$= .0985$

POINT	P	h	S	X
a	100	298.4	—	0
b	100	1187.2	1.6026	1.00
c	14.7	1046.2	1.6026	.893
d	14.7	48.02	—	0
e	100	48.2	—	—

6-12

(a) dewpoint: $\Sigma x_i = 1 = \Sigma y_i / k_i$

i	P=100		P=120		P=150	
	K_i	x_i	K_i	x_i	K_i	x_i
C_3H_8	4	.125	3.7	.135	3	.167
C_4H_{10}	2	.15	1.75	.171	1.4	.214
C_6H_{14}	.4	.5	.38	.526	.3	.667
Σ		.775		.832		1.048

(b) dew point pressure $\cong$ 150 psi

composition: 16.7% PROPANE
21.4% BUTANE
66.7% HEXANE

(c) bubble point $\Sigma y = 1 = k_i x$

i	P=300		P=350		P=400	
	K_i	y_i	K_i	y_i	K_i	y_i
C_3H_8	1.8	.9	1.70	.85	1.6	.8
C_4H_{10}	.85	.255	.80	.24	.76	.228
C_6H_{14}	.19	.038	.16	.032	.16	.032
Σ		1.193		1.122		1.06

bubble point pressure $\cong$ 400 psi

80 % propane
22.8% butane
3.2% hexane

6-13

(a)

$$k\tau = \frac{X_A}{C_{A_0}(1-X_A)(M-X_A)}$$

1st reactor: $M = 3/2 = 1.5$

$$\tau = \frac{V}{v_0} = \frac{2}{8} = .25$$

$$k = \frac{.6}{(.25)(2)(1-.6)(1.5-.6)}$$

$$k = 10/3 = 3.33 \ \frac{ft^3}{\text{lbmole-min}}$$

$C_A = C_{A_0}(1-X_A) = 2(1-.6) = .8$

$C_B = C_A + 1 = 1.8$

2nd reactor: $M = \frac{1.8}{.8} = \frac{9}{4} = 2.25$

$(C_{A_0})_2 = 0.8$

$$k\tau(C_{A_0})_2 = 3.33(.25)(.8) = \frac{2}{3} = .667$$

$$.667 = \frac{X_{A_2}}{(1-X_{A_2})(2.25-X_{A_2})}$$

Reduces to

$$4X_{A_2}^2 - 19X_{A_2} + 9 = 0$$

$$X_{A_2} = \frac{19-\sqrt{19^2-4(4)(9)}}{8}$$

$X_{A_2} = 0.5336$

$$X_A = \frac{X_{A_1}C_{A_0} + X_{A_2}C_{A_1}}{C_{A_0}} = \frac{.6(2) + X_{A_2}(.8)}{2}$$

$X_A = .81344$

(b) $C_A = 2(1-X_A) = .3731 \ \frac{\text{lbmole}}{ft^3}$

$C_B = C_A + 1 \quad = 1.3731$ lbmole/ft³

$C_R = 2 - C_A \quad = 1.6269$ lbmole/ft³

$C_S = C_R \quad = 1.6269$ lbmole/ft³

6-14

(a)

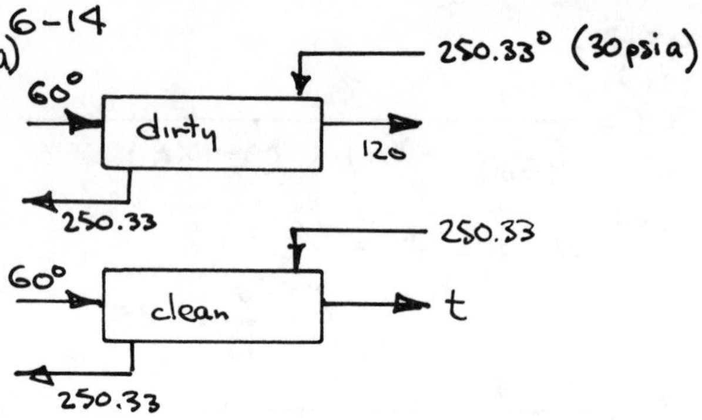

$Q_{DIRTY} = mc_p \Delta t$

$= (125 \times 8.3373)(1)(120-60)$

$= 62530$ Btu/min

$$\Delta t_{lm} = \frac{(250.33-60)-(250.33-120)}{\ln \frac{190.33}{130.33}} = 158.44°$$

$Q_{DIRTY} = UA\,t_{lm} = 220\,A\,(158.4)$

$A = 1.794\ ft^2$

Clean:

$$U_{DIRTY} = \frac{1}{\frac{1}{U_{CLEAN}} + .001} = 220$$

$U_{CLEAN} = 282\ Btu/hr\,ft^2\,°F$

Assume:

$Q_{CLEAN} = Q_{DIRTY}$

$62530 = U_{CLEAN}\,A\,\Delta t_{lm}$

$= 282(1.794)\,\Delta t_{lm}$

$$\Delta t_{lm} = 123.6 = \frac{190.33 - X}{\ln \frac{190.33}{X}}$$

trial & error yields

$X = 74.67$

$X = 250.33 - t$

$t = 175.66\ °F$

(b) when clean the steam pressure should be reduced to get 130°F outlet

6-14 (con't)

$$\Delta t_{lm} = \frac{(t_{STEAM}-60)-(t_{STEAM}-130)}{\ln \frac{t_{STEAM}-60}{t_{STEAM}-130}} = 123.6$$

$$123.6 = \frac{70}{\ln \frac{t-60}{t-130}}$$

$$\frac{t-60}{t-130} = 1.7618 \quad t_{STEAM} = 222°F$$

$\simeq 18$ psia

At startup reduce pressure to $\simeq 18$ psia and gradually increase to compensate for dirt factor

6-15

Assume:
1. fluid is water @ 70°F
2. pipe is steel $\varepsilon = .00015$
3. subscript "6-8" means going from 6" to 8", etc
4. teeR means run tee, teeB means branch tee
5. the tee at 6,8,10 is 10 inch, at 8,10,12 is 12 inch
6. Subscripts are nominal pipe sizes

$d_6 = 6.065''$ $d_8 = 7.981$ $d_{10} = 10.02$ $d_{12} = 11.938$

$D_6 = .5054$ ft $D_8 = .6651$ $D_{10} = .8350$ $D_{12} = .9965$

$L_6 = 200$ $L_8 = 450$ $L_{10} = 450$ $L_{12} = 500$

Sacham equation

$$f = \left\{-2\log\left(\frac{\varepsilon/D}{3.7} - \frac{5.02}{N_{Re}}\left[\log\frac{\varepsilon/D}{3.7} + \frac{14.5}{N_{Re}}\right]\right)\right\}^{-2}$$

$$N_{Re} = 50.6\,\frac{Q\rho}{d\mu} \quad \rho = 62.305 \quad \mu = .9768$$

6-15 (con't)

fittings & valves:

$\left(\frac{L}{D}\right)_{GLOBE} = 340$ $\left(\frac{L}{D}\right)_{ELL} = 30$

$\left(\frac{L}{D}\right)_{teeR} = 20$ $\left(\frac{L}{D}\right)_{teeB} = 60$

Expansion/contraction

$$K = f\frac{L}{D} = \left(1 + \frac{d_i^2}{d_2^2}\right)^2$$

$K_{6-10} = 0.4015$

$K_{10-8} = 0.1336$

$K_{10-12} = 0.0873$

$K_{8-12} = 0.3059$

$Q_6 = 2500$ gpm

$Q_{10} = 2500 - Q_8$

$Q_{12} = 2500$ gpm

$$h_f = 0.0311 \frac{fLQ^2}{d^5}$$

for $Q_A = 1907$

$(N_{Re})_A = 5\times10^5$ $(N_{Re})_B = 2\times10^5$

$\left(\frac{\varepsilon}{D}\right)_A = .0008$ $\left(\frac{\varepsilon}{D}\right)_B = .0003$

$f_A = .023$ $f_B = .025$

2nd iteration

$$Q_A^2 = 10.33\frac{f_B}{f_A}(2500 - Q_A)^2$$

$Q_A = 1925$ gpm

$Q_B = 575$

$f_A = .023$ $f_B = .025$ o.k.

The Bernoulli equation from tee to tee for leg A & B

6-15 (con't)

$\delta h_z + \delta h_p + \delta h_v + h_f = 0$ for each leg

$\therefore$ since $(\delta h_z)_A = (\delta h_z)_B$

$(\delta h_p)_A = (\delta h_p)_B$

$(\delta h_v)_A = (\delta h_v)_B$

$\therefore h_{fA} = h_{fB}$

$$\therefore Q_A^2 = \frac{f_B L_B d_A^5}{f_A L_A d_B^5}(2500 - Q_A)^2$$

Assume $f_A = f_B$ and correct later

$$Q_A^2 = K(2500 - Q_A)^2$$

$$K = \frac{L_B d_A^5}{L_A d_B^5}$$

$$L_A = D_A\left[\left(\frac{L}{D}\right)_{ell} + \left(\frac{L}{D}\right)_{globe} + \frac{L_{10}}{D_{10}}\right]$$

$L_A = 759$ ft

similarly

$L_B = 697$ ft

$\therefore K = 2.861$

$$Q_A = \frac{2500\sqrt{K}}{1 + \sqrt{K}}$$ solved for Q_A

$Q_A = 1571$ gpm

So far:

d	Q	N_{Re}	f
10.02	1571	506070	0.01525
7.981	929	375640	0.01609

correct for friction factor

$K = K_{OLD}\, f_B/f_A = 3.019$

$\therefore Q_A = 1587$ gpm

now:

d	Q	N_{Re}	f	h_f
10.02	1587	511111	.01524	8.964
7.981	913	369308	.01612	8.988

(6-15) con't

For whole system

$$\delta h_z + \delta h_p + \delta h_v + h_f = 0$$

$$\delta h_z = 150$$

$$Le_6 = L_6 + \left[\left(\frac{L}{D}\right)_{GLOBE} + \left(\frac{L}{D}\right)_{ell} + \frac{K_{6-10}}{f_6}\right] D_6$$

$$Le_8 = L_8 + \left[\left(\frac{L}{D}\right)_{globe} + \left(\frac{L}{D}\right)_{ell} + \left(\frac{L}{D}\right)_{teeB} + \frac{K_{10-8} + K_{8-}}{f_8}\right] D_8$$

$$Le_{10} = L_{10} + \left[\left(\frac{L}{D}\right)_{globe} + \left(\frac{L}{D}\right)_{ell} + \left(\frac{L}{D}\right)_{teeR} + \frac{K_{10-12}}{f_{10}}\right] D_{10}$$

$$Le_{12} = D_{12}\left[\left(\frac{L}{D}\right)_{teeR} + \left(\frac{L}{D}\right)_{teeB}\right] + L_{12}$$

$$v = 0.408 \frac{Q}{d^2}$$

i	N_{Re_i}	Le_i	f_i	h_{f_i}
6	1330384	400.1	0.0155	146.60
10	511184	780.4	.0152	9.22
8	369217	754.1	.0161	9.73
12	675890	579.7	.0145	6.768

$v_6 = 27.73$ ft/sec

$v_{12} = 7.16$ ft/sec

using circuit through 10-in pipe

$$\delta h_p = -\delta h_z - \frac{v_{12}^2 - v_6^2}{2g} - h_{f_6} - h_{f_{10}} - h_{f_{12}}$$

$\delta h_p = -301.44$ ft

$$\delta h_p = \frac{(P_2 - P_1)}{\rho} 144 \qquad P_2 = 50$$

$P_1 = 180.5$ psig

NOTE: ANSWER COULD VARY BY 1% DEPENDING ON WHICH LINE SUDDEN EXPANSION/CONTRACTION IS TAKEN.

6-16 IDEAL GAS

$$C_{A_0} = \frac{P_{A_0}}{RT} = \frac{340/760}{.0821(273+270.1)}$$

$= .0100$ mole/liter

$$\varepsilon_A = \frac{2-1}{1} = 1$$

$$V = V_0(1 + \varepsilon_A X_A)$$

$$P = P_0(1 + \varepsilon_A X_A)$$

$$X_A = \left[\frac{P}{P_0} - 1\right] / \varepsilon_A$$

first order

$$k_1 = -\ln(1 - X_A)/t$$

2nd order

$$k_2 = \left[\frac{(1+\varepsilon_A)X_A}{1 - X_A} + \varepsilon_A \ln(1 - X_A)\right] / C_{A_0} t$$

P	t	P/P_0	X_A	k_1	k_2
340	0	1	0	–	–
357	.36	1.05	.05	.1425	14.99
391	1.26	1.15	.15	.12898	15.11
408	1.84	1.20	.2	.12127	15.05
425	2.52	1.25	.25	.11416	15.04
459	4.30	1.35	.35	.10018	15.03
475	5.48	1.397	.397	.0923	14.80
510	8.71	1.5	.5	.07958	15.0
595	30.75	1.75	.75	.04508	15.0
663	233.36	1.95	.95	.01284	15.0

$\bar{k} = 15.002$ 2nd order

liter/mole-min

6-17

4" sched 40 ID = 4.026"

OD = 4.5"

K = 0.037 Btu/hr ft °F (85% magnesia)

K = 27 Btu/hr ft °F (cast iron)

$$q = \frac{\pi(1)\Delta t}{\frac{1}{h_i D_i} + \frac{\ln(D_o/D_i)}{2K_{pipe}} + \frac{1}{h_o D_o} + \frac{\ln D_{ins}/D_o}{2K_{ins}} + \frac{1}{h_{ins} D_{ins}}}$$

$$= \frac{\pi(1)(500-70)}{\frac{1}{50\left(\frac{4.026}{12}\right)} + \frac{\ln(4.5/4.026)}{2(27)} + \frac{\ln(6.5/4.5)}{2(.037)} + \frac{1}{5\left(\frac{6.5}{12}\right)}}$$

$q = 250.15$ Btu/hr-ft

6-18

From plot on next page

Chlorobenzene req'd = 8349 lb.

6-19

$\Delta H_{rxn} = \Delta H_{RXN\,25°C} + \Delta H_{REACTANTS} + \Delta H_{PRODUCT}$

$$\Delta H_{rxn} = \int_{773}^{298} C_p dT + \Delta H_{RXN,25°C} + \int_{298}^{773} C_p dT$$

$$\Delta H_{REACTANTS} = 32\left[5.34(298-773) + \frac{.0115}{2}(298^2 - 773^2)\right] + 71\left[8.28(298-773) + \frac{.00056}{2}(298^2 - 773^2)\right]$$

$= -464130$ cal

$$\Delta H_{PROD} = 101\left[6.26(773-298) + \frac{.0094}{2}(773^2 - 298^2)\right] + 73.0\left[6.70(773-298) + \frac{.00084}{2}(773^2 - 298^2)\right]$$

$= 789735$ Cal

6-19 (con't)

$\Delta H_{RXN,25} = -19600(2) - 22063(2) - [17899(2) + 0]$

$= -47525$

$q = -464130 - 47578 + 789735$

$= +278077 \frac{\text{cal}}{\text{2 moles } CH_4}$

$= 139039$ Cal/mole CH_4

[b] heat is added [+]

6-20

Use PW method

Let t = tax rate

Rental:

$PW_R = (-\$.28/\text{mile})(12000 \text{ miles})(1-t)(P/A, 15\%, 4)$

$= (-.28)(12000)(1-.4)(2.8550)$

$= -\$5756$

Buy $D_1 = 9000(.25)(t)$; $D_2 = 9000(.38)t$

$D_3 = 9000(.37)t$

$PW_B =$

$9000\{-1 + t[.25(P/F, 15\%, 1) + .37(P/F, 15\%, 2) + .38(P/F, 15\%, 3)]\}$

$+ 2000(P/F, 15\%, 4) - 800(P/A, 15\%, 4) \times (1-t)$

$PW_B = 9000[-1 + .4(.25(.8696) + .37(.7561) + .38(.6575))] + 2000(.5718) - 800(2.8550)(1-.4)$

$PW_B = -\$6538$

Rent because PW has lower cost.

Problem 6–18

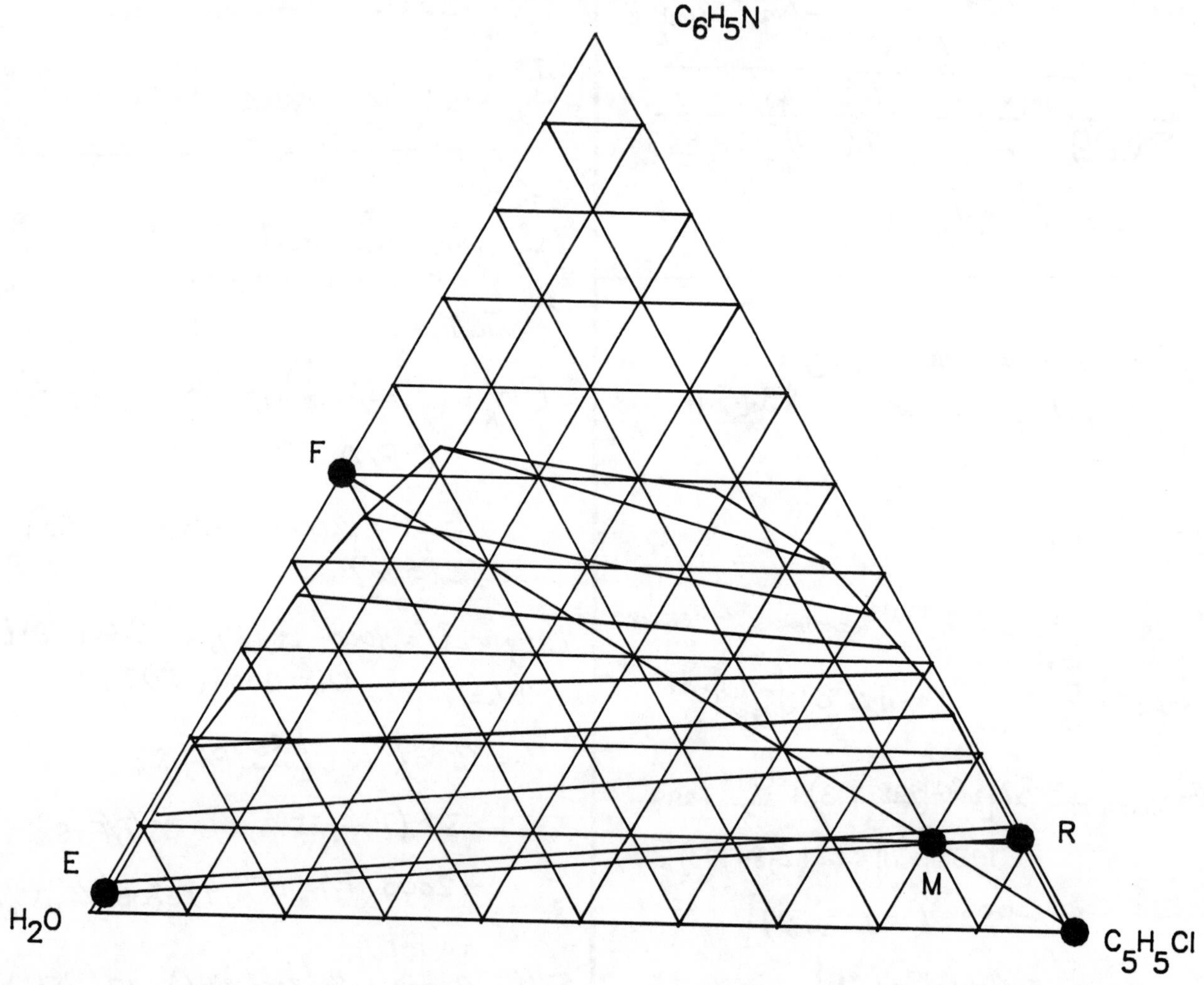

7-1

data N_2, $T[\equiv °K]$, $C_p[\equiv cal/gmole°K]$

Perry $\quad C_p = 6.5 + .001T$

Lewis & Randall $C_p = 6.83 + .0009T - \frac{12000}{T^2}$

Hougan & Watson

$$C_p = 6.903 - 0.0003753T + 0.193\times10^{-5}T^2 - 0.6861\times10^{-9}T^3$$

$$dH = \left(\frac{\partial H}{\partial T}\right)_P dT + \left(\frac{\partial H}{\partial P}\right)_T dP;$$

$$\left(\frac{\partial H}{\partial T}\right)_P = C_P$$

$$\left(\frac{\partial H}{\partial P}\right)_T = -T\left(\frac{\partial V}{\partial T}\right)_P + V$$

for ideal gas $\left(\frac{\partial H}{\partial P}\right)_T = 0$

real gas $\left(\frac{\partial H}{\partial P}\right)_T = -\frac{RT^2}{P}\left(\frac{\partial Z}{\partial T}\right)_P$

1st approx.: N_2 ideal @ 25°C, 1atm
ideal @ 700°, 25 atm

$$\Delta H = \int_{298}^{973} C_p dT = \int_{298}^{973}\left[a + bT + cT^2 + dT^3 + \frac{e}{T^2}\right]dT$$

$$\Delta H = a(973-298) + \frac{b}{2}(973^2 - 298^2) + \frac{c}{3}(973^3 - 298^3) + \frac{d}{4}(973^4 - 298^4) - e\left(\frac{1}{973} - \frac{1}{298}\right)$$

$$\Delta H = 675a + 428963b + 2.9823\times10^8 c + 2.221\times10^{11} d + 0.002328e$$

Perry:

$$\Delta H = 675(6.5) + .001(428963) = 4816.5 \frac{cal}{gmole}$$

7-1 (con't)

L&R:

$$\Delta H = 675(6.83) + 428963(.0009) - 002328(-12000)$$
$$= \mathbf{4968.4}\ cal/gmole$$

H&W

$$\Delta H = 675(6.903) + 428963(-.0003753) + 2.9823\times10^8(.193\times10^{-5}) + 2.221\times10^{11}(-.686\times10^{-9})$$
$$= \mathbf{4921.7}\ cal/gmole$$

2nd approx:

N_2 ideal at 25°C real at 700°C

$$\Delta H = C_p dT + \left[V - T\left(\frac{\partial V}{\partial T}\right)_P\right]dP$$

$$= C_p dT - \int_1^{25} RT^2\left(\frac{\partial Z}{\partial T}\right)_P \frac{dP}{P}$$

data for Z: N_2

T°K	20atm	40atm	25atm
800	1.0084	1.0157	**1.0096**
1000	1.0074	1.0136	**1.0085**

looks like N_2 is nearly ideal at 25 atm

$$\left(\frac{\partial Z}{\partial T}\right)_P \cong \frac{\Delta Z}{\Delta T} = \frac{1.0085 - 1.0096}{1000 - 800}$$
$$= -5.5\times10^{-6}\ °K^{-1}$$

$$\left(\frac{\partial H}{\partial P}\right)_{973} \cong -RT^2\left(\frac{\partial Z}{\partial T}\right)_P \int_1^{25}\frac{dP}{P}$$

$$= -RT^2\frac{\Delta Z}{\Delta T}\left[\ln 25 - \ln 1\right] \quad (\ln 1 = 0)$$

$$= -1.987(973)^2(-5.5\times10^{-6})\ln 25$$

$$= 33.3\ cal/gmole$$

this correction is less than the uncertainty for C_p.

7-2

ERGUN EQUATION

$$\frac{(-\Delta P) g_c D_P \varepsilon^3}{L \rho v_{sm}^2 (1-\varepsilon)} = 150 \frac{1-\varepsilon}{N_{Re}} + 1.75$$

$-\Delta P$ = pressure drop, psi

L = bed length, ft

ρ = fluid density

g_c = 32.2 (lb/lb$_f$)(ft/sec^2)

ε = bed porosity

N_{Re} = average Reynold's number based on particle diameter

D_{P}, = $D_P v_{sm} \rho / \mu$

D_P = particle diameter, ft

$D_P = 6V_P/S_P$ {V = particle volume, ft^3; S = particle surface area, ft^2}

$D_P = 6(1/4)^3/[6(1/4)^2] = 0.25$ in

$$\text{sphericity} = \psi = \frac{\pi (6V_P/\pi)^{2/3}}{A_P}$$

$$\psi = \frac{\pi \left(\frac{6(1/4)^3}{\pi}\right)^{2/3}}{6(1/4)^2} = 0.806$$

$\varepsilon = 0.44$ from ψ vs. ε plot

"Principles of Unit Operations", Foust, et al.

Assuming ΔP is small & v_{sm} func. of temp.

$$\rho_{IN} = \frac{PM}{RT_1} = \frac{100(29)}{10.73(540)} = 0.5 \frac{\text{lb}}{\text{ft}^3}$$

$$\rho_{OUT} = \frac{100(29)}{10.73(860)} = 0.314 \frac{\text{lb}}{\text{ft}^3}$$

$$v_{sm} = \frac{1000/.5 + 1000/.314}{2} = 2592 \frac{\text{ft}}{\text{hr}}$$

$$\frac{N_{Re}}{1-\varepsilon} = \frac{D_P G/\mu}{1-\varepsilon} = \frac{0.25(1000)}{12(1-.44)(.023 \times 2.42)} = 668$$

$$\frac{-\Delta P g_c D_P \varepsilon^3}{L \rho v_{sm}^2 (1-\varepsilon)} = \frac{150}{668} + 1.75 = 1.974$$

$$-\Delta P = \frac{1.974 L \rho v_{sm}^2 (1-\varepsilon)}{g_c D_P \varepsilon^3} = 0.283 \text{ psi}$$

7-3

TDH = h_{pump} in Bernoulli Eqtn.

point 1: top of suction reservoir

point 2: pressure gauge

Bernoulli:

$$Z_1 + \frac{v_1^2}{2g}\!\!\!\!\!\!\!\!\!\!\!\!\! \rightarrow 0 + \frac{P_1}{\rho} \rightarrow 0 = Z_2 + \frac{v_2^2}{2g} + \frac{P_2}{\rho} + h_{pump} + h_f$$

$$5 = 60 + \frac{v_2^2}{2g} + \frac{300(144)}{60.80} + h_{pump} + h_f$$

v_2 estimated from Crane

150 gpm through 4" sched 40

@ 60°F = 1.67 ft/sec

@ 170° v = 1.67 62.37/60.80

= 1.71 ft/sec

$$5 = 60 + \frac{v_2^2}{2g} + 710.5 + h_{pump} + h_f$$

$h_{pump} + h_f = -765.6$ ft

Equiv lengths suction

Elbows + check valve + gate valve

= 3(30) + 135 + 13 = 238 pipe diams

L = 20 + 238 (.5054) = 140.3 ft

discharge:

elbows + check valve + globe valves

3(30) + 135 + 2(340) = 905

L = 200 + 905 (.3355) = 503.6 ft

$h_f = h_{suction} + h_{disch}$

$$= \frac{140.3}{100}(1.8) + \frac{503.6}{100}(8.5)$$

= 45.3 ft

$h_{pump} = -765.6 - 45.3 = 810.9$ ft

7-4

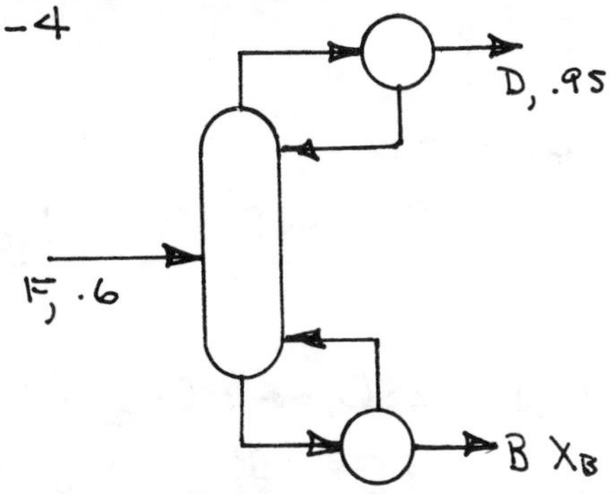

Mat bal: Basis 100 mole/hr feed

$100 = D + B$

$.6(100) = .95D + X_B B$

$$D = \frac{60 - 100X_B}{.95 - X_B}$$

MUST FIND X_B GRAPHICALLY

AT MIN REFLUX, RECT LINE INTERSECTS Q LINE AT EQUIL. LINE.

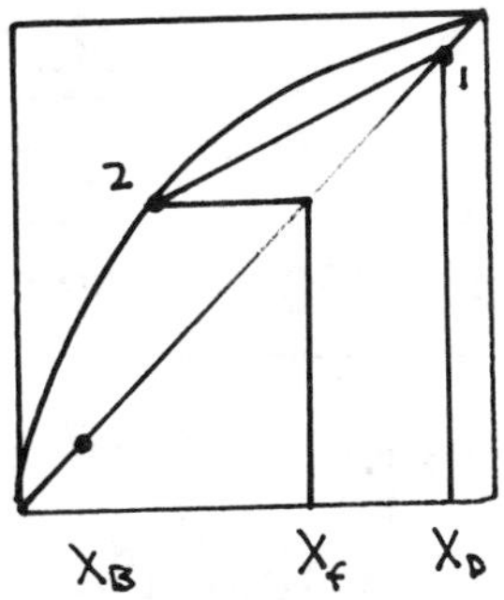

ENDPOINTS OF RECT LINE

(.95, .95) $(X^*, .6)$

$\alpha = 2.2$ $X^* = .405$

$$\therefore \left(\frac{L}{V}\right)_{min} = \frac{.95 - .6}{.95 - .405} = .642$$

$$\frac{R_m}{R_m + 1} = .642 \ ; \ R_m = 1.795$$

$R = 1.5(1.795) = 2.69$

$$\frac{L}{V} = \frac{2.69}{2.69 + 1} = 0.729$$

OP LINE $y = 0.729x + .2575$

7-4 (con't)

Q-line: $y = 0.6$

END POINTS OF RECT LINE

(.95, .95) & (X, .6)

$0.6 = 0.729x + .2575$

$x = .4698$

(.95, .95) & (.4698, .6)

Since the actual number of 50% effective plates = 16, then number of equilibrium plates = 8 + 1 for reboiler + 1 for condenser = 10

With number of plates known rectification line known, equil line known, use following procedure

1. Guess X_B, draw (calculate) strip line
2. Step off 10 stages from X_D (rect. section doesn't change)
3. If equil stages fall below X_B, decrease X_B, go to 2. If equil stages fall above X_B, increase X_B, go to 2. If equil stages fall on X_B calculate D.

Calculational method:

rect. eqtn: $y = \frac{L}{V}x + b$; $\frac{L}{V} = .72917$

equil line: $x = \frac{y}{\alpha + y(1 - \alpha)}$ $\alpha = 2.2$

Rect. section

i	x_i^*	y_i
1	.8962	.95
2	.8227	.9108
3	.7318	.8572
4	.6324	.7910
5	.4371	.7185
6	.4567	.6490

7-4 (con't)

Strip section: if $y_{10} = X_B$ guessed right X_B

① Guess X_B

② calculate slope: $\frac{L}{V} = \frac{.6 - X_B}{.47 - X_B}$

③ strip line: $y = \frac{L}{V}x + X_B\left(1 - \frac{L}{V}\right)$

i	$X_B = .1$; $\frac{L}{V} = 1.351$		$X_B = .15$; $\frac{L}{V} = 1.406$		$X_B = .125008$; $\frac{L}{V} = 1.376$	
	x	y	x	y	x	y
6	.5377	.6487	.5368	.6487	.5368	.6487
6	.4563	.6487	.4563	.6487	.4563	.6487
7	.4563	.5815	.4563	.5808	.4563	.5812
7	.3871	.5185	.3864	.5808	.3868	.5812
8	.3871	.4880	.3864	.4824	.3868	.4854
8	.3023	.4880	.2976	.4824	.3000	.4854
9	.3023	.3733	.2976	.3575	.3000	.3661
9	.2131	.3733	.2019	.3575	.2079	.3661
10	.2131	.2528	.2019	.2230	.2079	.2391
10	.1335	.2528	.1153	.2230	.12500	.2391
		(.1450)	.1153	(.1013)		(.1250)

$X_B = .125008$

$$D = \frac{60 - 100(.125008)}{.95 - .125008}$$

$D = 57.23$ lbmoles/hr

for every 100 lbmoles feed.

7-5

(a) Stoichiometric

$$C_8H_{18} + 12.5O_2 \rightarrow 8CO_2 + 9H_2O$$

100% excess

$$25\left(\frac{79}{21}\right)N_2 + C_8H_{18} + 25O_2 \rightarrow 8CO_2 + 9H_2O + 25\left(\frac{79}{21}\right)N_2 + 12.5O_2$$

assume 1 mole C_8H_{18}

	moles REACTANTS	moles PRODUCTS	mole fract PRODUCTS
C_8H_{18}	1	0	0
N_2	94	94	.7611
O_2	25	12.5	.1012
CO_2	0	8	.0648
H_2O	0	9	.0728
total	120	123.5	.9999

(b) $P_{H_2O} = .0728(14.7) = 1.0702$ psia

$t_{sat\ @\ 1.07\ psia} = 104.1°F$ (dew point)

7-6 For 2 CSTR's in series

(a) with = volumes 1st order (subscript = temperature °C)

$$\frac{C_{A_2}}{C_{A_0}} = \frac{1}{[1 + K_{30}\tau_{30}]^2}$$

if yield is same at both temperatures

$$\frac{1}{[1+K_{30}\tau_{30}]^2} = \frac{1}{[1+K_{70}\tau_{70}]^2}; \quad K_{30}\tau_{30} = K_{70}\tau_{70}$$

$$\frac{\tau_{30}}{\tau_{70}} = \frac{K_{70}}{K_{30}} = \frac{1230e^{-6000/343}}{1230e^{-6000/303}} = 10.07$$

PROBLEM 7-4

McCabe–Thiele Plot

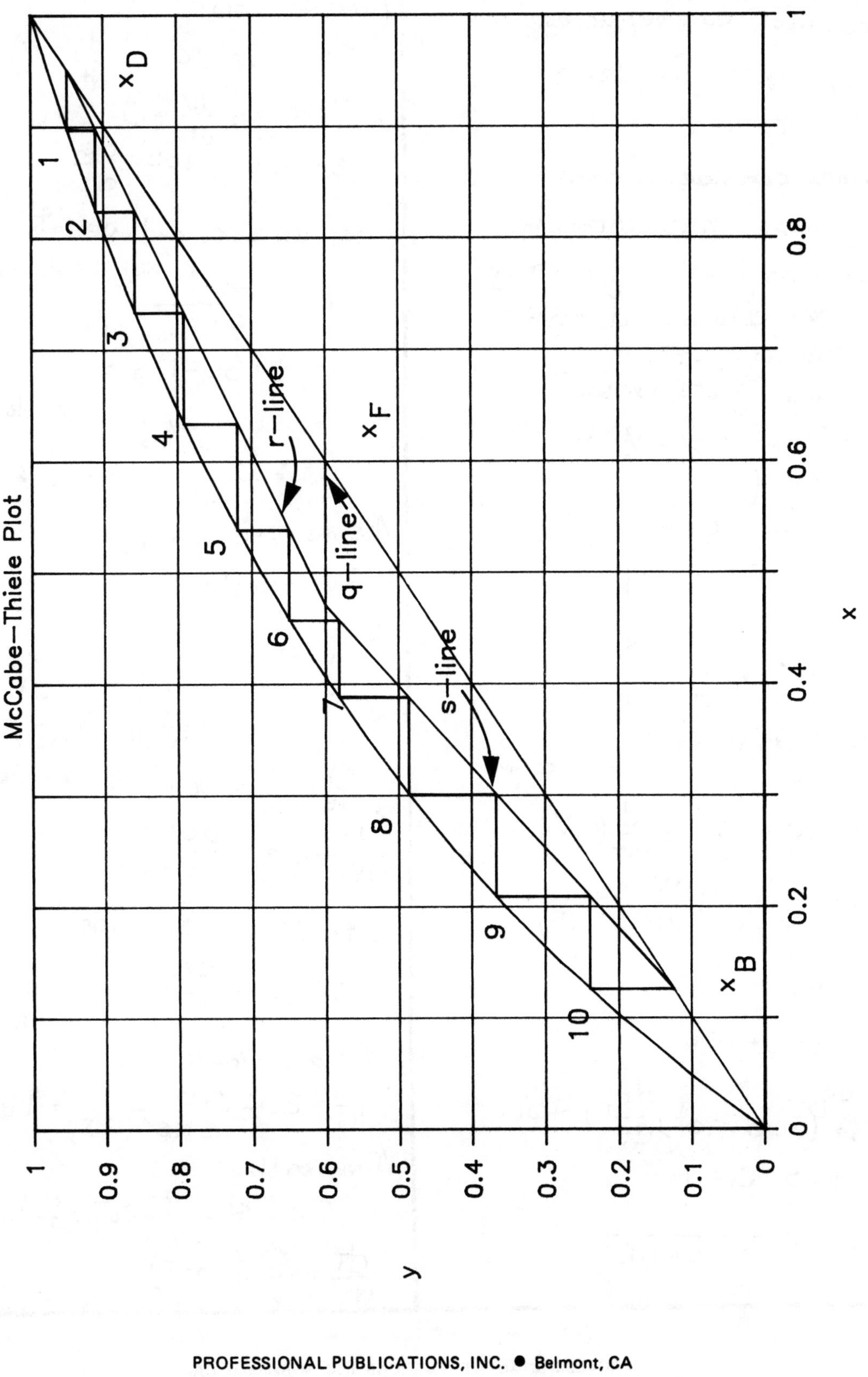

7-6 (con't)

$\tau_{30} = V/\upsilon_0 = 1000/100 = 10 \text{ min}$

$\tau_{70} = 10/10.05 = 0.993 \text{ min}$

throughput can be increased 10.07 times for same yield at 70°C.

(b) Other considerations:

1. Pumps sized correctly.
2. Pumps handle higher temp?
3. Mixer efficient at higher throughput?
4. Same yield needed? Would higher yield be beneficial?

7-7

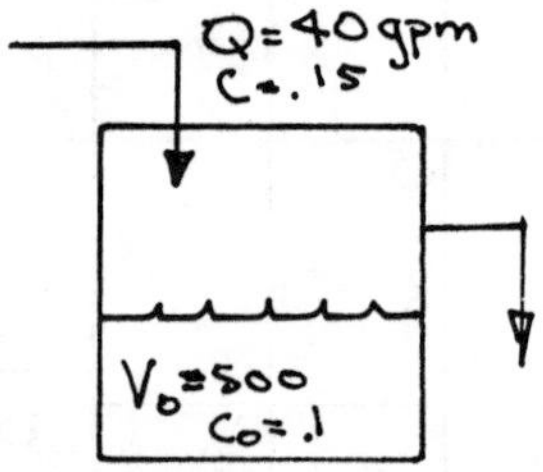

before overflow:

In − Out = accum

$$40(.15) - 0 = \frac{d}{dt}(C_T V)$$

$$6 = V\frac{dC_T}{dV} + C_T\frac{dV}{dt}$$

$$\frac{dV}{dt} = 40$$

$$V = 500 + 40t$$

diff. eq

$$6 = (500 + 40t)\frac{dC_T}{dt} + 40C_T$$

B.C.: $t=0$ $C_T = .1$

$$C_T = .15 - \frac{1}{20 + 1.6t}$$

7-7 (con't)

at overflow $t = 12.5$

$$C_T = .15 - \frac{1}{20 + 1.6(12.5)} = .125$$

After overflow:

$$40(.15) - 40C_T = \frac{d}{dt}(C_T V)$$

$$V = 1000; \quad \frac{dV}{dt} = 0$$

$$6 - 40C_T = 1000\frac{dC_T}{dt}$$

sep variables and integrate

BC: $t=0$ $C_T = .125$

$$C_T = \frac{6 - e^{-t/25}}{40}$$

at $t = 30 - 12.5 = 17.5$ min overflow

$$C_T = \frac{6 - e^{-17.5/25}}{40} = .1376 \text{ or } 13.76\%$$

Assume $C_p = C_{p,15\%} = C_{p,10\%}$

before overflow:

$$QC_p\,180 - 0 = \frac{d}{dt}(C_p V T)$$

$$= C_p\left(T\frac{dV}{dt} + V\frac{dT}{dt}\right)$$

$\frac{dV}{dt} = Q$ and $V = 500 + Qt$

$$\int\frac{dT}{180 - T} = \int\frac{Q\,dt}{500 + Qt}$$

with BC $t=0$ $T=100$

$$T = 180 - \frac{40}{.5 + .04t} \quad \text{at overflow}$$

At overflow:

$$T = 180 - \frac{40}{.5 + .04(12.5)} = 140°\text{F}$$

After overflow:

$$QC_p\,180 - QC_pT = \frac{d}{dt}(C_p V T)$$

$$\frac{dT}{dt} = \frac{Q}{V}(180 - T)$$

7-7 (con't)

With B.C. $t=0 \quad T=140$

$$T = 180 - 40e^{-Qt/V}$$

at $t = 30 - 12.5 = 17.5$ min

$$T = 180 - 40e^{-40(17.5)/1000}$$

$$T = 161.14°F$$

7-8

Reliability:

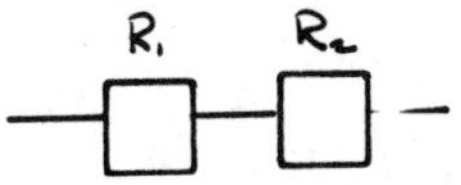

OVERALL RELIABILITY

$R = R_1 R_2$

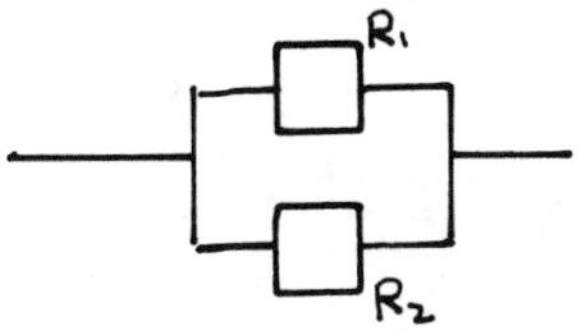

$R = R_1 + R_2 - R_1 R_2$

To add a new unit in parallel to system with Reliability R_N

$$R_{N+1} = R_N + R_{NEW} - R_N R_{NEW}$$

Maint cost $= 1000(24)(365)(1 - R_{OVERALL})$

Total cost = cap cost + maint cost

Consider only 2 possibilities

(1) replace all pumps with N new ones assume 2 single stage pumps in series equals one double stage.

(2) add N pumps to old system

Thus, a total of 4 scenarios: possibility (1) using single stage pumps in pairs or double stage pumps Or (2) with single stage pairs or double stage pumps.

7-8

N Pairs of single stage pumps in parallel $R_{PAIR} = .75(.75) = .5625$

N	$R_{overall}$	MAINT$	CAP$	TOTAL$
1	.5625	3832.5	25	3857.5
2	.8086	1677.6	50	1727.6
3	.9163	732.2	75	808.2
4	.9634	370.6	100	420.6
5	.9840	140.1	125	265.2
6	.9930	61.3	150	211.3
7	.9969	27.2	175	203.2 ←
8	.9986	12.2	200	212.2

BEST ARRANGEMENT FOR PAIRS OF SINGLE STAGE IS 7 IN PARALLEL

N DOUBLE STAGE PUMPS IN PARALLEL $R = 0.5$

N	$R_{OVERALL}$	MAINT$	CAP$	TOTAL$
1	.5	4380	20	4400
2	.75	2190	40	2230
3	.875	1095	60	1155
4	.9375	547	80	627
5	.9688	273.3	100	373.3
6	.9844	136.7	120	256.7
7	.9922	68.3	140	208.3
8	.9961	34.2	160	194.2 ←
9	.9980	17.5	180	197.5

BEST ARRANGEMENT OF DOUBLE PUMPS IS 8 IN PARALLEL

7-8 (con't)

EXISTING SYSTEM:
3 DOUBLE STAGE PUMPS IN PARALLEL HAS RELIABILITY R= 0.875

CONSIDER ADDING N PAIRS OF SINGLE STAGE PUMPS IN PARALLEL WITH EXISTING SYSTEM

N	$R_{overall}$	MAINT$	CAP$	TOTAL$
0	.875	1095	0	1095
1	.9453	479	25	504
2	.9761	209	50	259
3	.9895	92.0	75	167.0
4	.9954	40.3	100	140.3 ←
5	.9980	17.5	125	142.5

OPTIMUM: ADD 4 pairs single stage pumps IN PARALLEL WITH EXISTING SYSTEM

CONSIDER ADDING N DOUBLE STAGE PUMPS IN PARALLEL

N	$R_{OVERALL}$	MAINT$	CAP$K	TOTAL$K
0	.875	1095	0	1095
1	.9375	548	20	568
2	.9688	273	40	313
3	.9844	137	60	197
4	.9922	68.3	80	148.3
5	.9961	34.2	100	134.2 ←
6	.9980	17.5	120	137.5

7-8 (con't)

OPTIMUM AND LOWEST COST ARRANGEMENT:
ADD 5 DOUBLE STAGE PUMPS IN PARALLEL WITH EXISTING SYSTEM

7-9

Heat loss by natural convection and radiation

$$Q = \sigma \varepsilon A\left[\left(\frac{T_2}{100}\right)^4 - \left(\frac{T_1}{100}\right)^4\right] + hA\Delta t$$

to use correlations compute:

$$N_{Gr} = \frac{gL^3 \beta \Delta t\, \rho^2}{\mu^2}$$

$$N_{Pr} = \frac{C_p \mu}{k}$$

$t_{STEAM} = 274°$ $t_{ROOM} = 70°$

$\varepsilon = 0.8$

$L = 100$ ft; $\beta = 3671.6$; $\Delta t = 204$

$\rho = 0.07528$ lb/ft³; $k = .015$ Btu/hr-ft °F

$\mu = .018 \times 2.42$ lb/hr-ft

$N_{Gr} = 7.35 \times 10^{13}$ $N_{Pr} = .741$

$N_{Gr} N_{Pr} = 5.14 \times 10^{13} > 10^9$

$\therefore h_{air} = 0.18 \Delta t^{.33} = 0.18(208)^{.33}$

$h_{air} = 1.048$ Btu/hr-ft²°F

$$Q = .173(.8)\left(\frac{100\pi}{4} \cdot \frac{6.625}{12}\right)^2\left[\left(\frac{274+460}{100}\right)^4 - \left(\frac{70+460}{100}\right)^4\right] + 1.048(100\pi)\left(\frac{6.625}{12}\right)(274-70)$$

$Q = 87815$ Btu/hr

7-9 (con't)

$h_{fg} \cong 928.75$ Btu/lb @ 274 °F (steam tables)

$W = Q/h_{fg} = 87815/928.75$

$= 94.6$ lb/hr condensate formed

7-10

cash flow

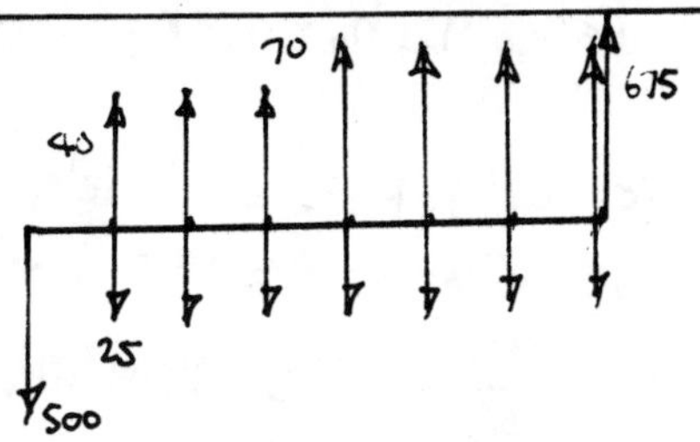

revenues simplified to:

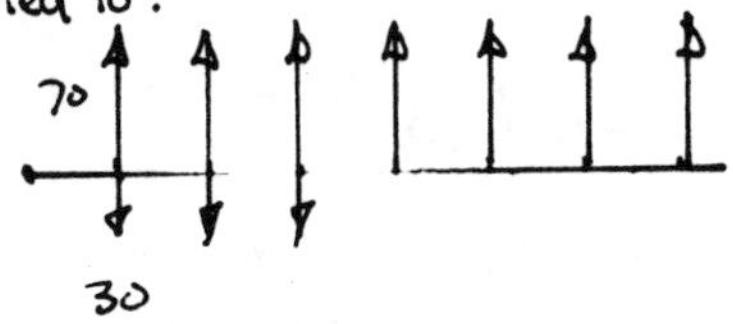

RULES:

1. NO TAX ON INITIAL INVESTMENT
2. NO TAX ON RESALE FROM B.V TO 0.
3. INITIAL COST EXCEEDED ON RESALE TAXED AT 40% (CAP GAINS)
4. REVENUE & COST AFTER TAX VALUE IS OBTAINED BY MULTIPLYING BY (1-t)
5. DEPRECIATION AFTER TAX VALUE IS OBTAINED BY MULTIPLYING BY t

S.L. Depreciation $= 500/15 = \$33.3$ k/yr.

BV after 7 years $= 500 - 7(33.3) = \$266.7$

$PW = -500 + 33.3(P/A, i\%, 7)(.4) +$
$-25(P/A, i\%, 7)(1-.4)$
$+70(P/A, i\%, 7)(1-.4)$
$-30(P/A, i\%, 3)(1-.4)$
$+266.7(P/F, i\%, 7) + (500-266.7)(P/F, i\%, 7)(1-.4)$
$+(675-500)(P/F, i\%, 7)(1-.6)(.4)$
$+(675-500)(P/F, i\%, 7)(.6)$

7-10 (con't)

$PW = -500 + 40.32(P/A, i\%, 7)$
$-18(P/A, i\%, 3) + 539.7(P/F, i\%, 7)$

set PW=0

trial & error

i	PW
7	-25.348
6	+ 3.809
6.2	- 2.19
6.1267	- 0.00196

RoR = 6.126%

7-11

(1) $[H^+][H_2PO_4^-] = K_1[H_3PO_4]$

(2) $[H^+][HPO_4^=] = K_2[H_2PO_4^-]$

(3) $[H^+][PO_4^{\equiv}] = K_3[HPO_4^=]$

(4) $[Na^+] = 0.15$

(5) $[HPO_4^=] + [H_2PO_4^-] + [PO_4^{\equiv}] + [H_3PO_4] = 0.1$

charge bal.

(6) $[H^+] + [Na^+] = 2[HPO_4^=] + [OH^-] + 3[PO_4^{\equiv}] + [H_2PO_4^-]$

(7) $[H^+][OH^-] = K_w$

put 4 → 6 & add 5 & 6 to get proton condition

$[H^+] + [H_3PO_4] + .05 = [OH^-] + [HPO_4^=] + 2[PO_4^{\equiv}]$

If pH is near 7, then from the value of K_1 & $[H^+]$ in (1) results in $[H_3PO_4]$ is negligible compared to $[H_2PO_4^-]$

From the value of K_3 & $[H^+]$ in (3) $[PO_4^{\equiv}]$ is negligible compared to $[HPO_4^=]$.

$[OH^-]$ & $[H^+]$ are negligible compared to $[HPO_4^=]$ & $[H_2PO_4^-]$

$[HPO_4^=] \cong 0.05$

then $H_2PO_4^- \cong 0.05$

then $[H^+][0.05] = K_2[.05]$; $[H^+] = 6.15 \times 10^{-8}$

7-11 (con't)

pH = 7.21

check assumptions

(1) $6.15\times10^{-8}[.05] = 5.9\times10^{-3}[H_3PO_4]$

$[H_3PO_4] = 5.21\times10^{-7}$

(3) $6.15\times10^{-8}[PO_4^{\equiv}] = 4.81\times10^{-13}(.05)$

$[PO_4^{\equiv}] = 3.9\times10^{-7}$

Assumptions OK,

pH = 7.21

7-12

1st order PFR

$$k\tau = -(1+\varepsilon_A)\ln(1-X_A) - \varepsilon_A X_A$$

$$\varepsilon_A = \frac{V_{x=1} - V_{x=0}}{V_{x=0}} = \frac{8-5}{5} = \frac{3}{5} = 0.600$$

$$\tau = \frac{-(1+.6)\ln(1-.8) - .6(.8)}{10}$$

$\tau = 0.2095$ hr

PV = nRT

$$v_0 = \frac{nRT}{P} = \frac{5\,\frac{\text{lbmoles}}{\text{hr}}\ .73023\,\frac{\text{atm-ft}^3}{\text{lbmole-°R}}(1200+460)\ \text{°R}}{4.6\ \text{atm}}$$

$v_0 = 1317.6$ ft^3/hr

$V = \tau v_0 = 1317.6(.2095)$

$V = 276$ ft^3

7-13

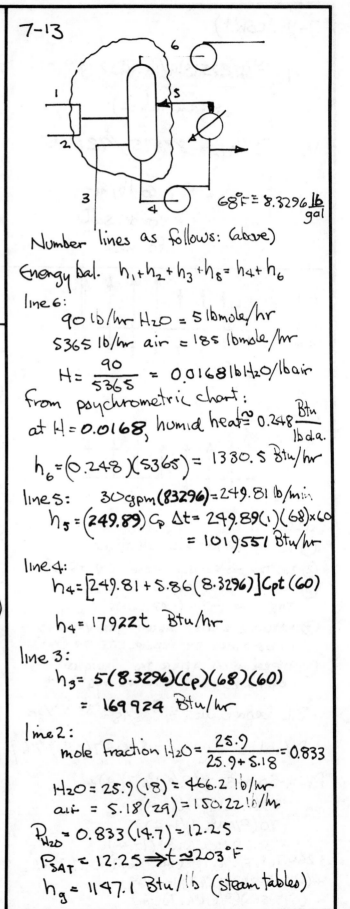

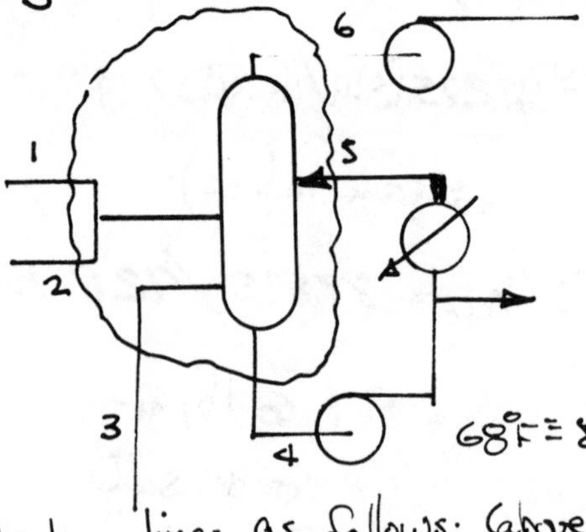

68°F = 8.3296 lb/gal

Number lines as follows: (above)

Energy bal. $h_1 + h_2 + h_3 + h_5 = h_4 + h_6$

line 6:

90 lb/hr H_2O = 5 lbmole/hr

5365 lb/hr air = 185 lbmole/hr

$$H = \frac{90}{5365} = 0.0168\ \text{lb}\,H_2O/\text{lb air}$$

from psychrometric chart:

at H = 0.0168, humid heat ≈ 0.248 $\frac{\text{Btu}}{\text{lb d.a.}}$

$h_6 = (0.248)(5365) = 1330.5$ Btu/hr

line 5: 30 gpm (8.3296) = 249.81 lb/min

$h_5 = (249.89)\,C_p\,\Delta t = 249.89(1)(68)\times60$

$= 1019551$ Btu/hr

line 4:

$h_4 = [249.81 + 5.86(8.3296)]C_p t(60)$

$h_4 = 17922t$ Btu/hr

line 3:

$h_3 = 5(8.3296)(C_p)(68)(60)$

$= 169924$ Btu/hr

line 2:

$$\text{mole fraction } H_2O = \frac{25.9}{25.9 + 5.18} = 0.833$$

$H_2O = 25.9(18) = 466.2$ lb/hr

air $= 5.18(29) = 150.22$ lb/hr

$P_{H_2O} = 0.833(14.7) = 12.25$

$P_{SAT} = 12.25 \Rightarrow t \cong 203°F$

$h_g = 1147.1$ Btu/lb (steam tables)

7-13 (con't)

air: from table 158.9 Btu/lb

$h_2 = 1147.1(466.2) + 158.9(150.22)$
$= 558648$ Btu/hr

line 1: humid heat 0.245 Btu/lb

$h_1 = 0.245(5220) = 1278.9$ Btu/hr

$\therefore$ $1278.9 + 558648 + 169924 + 1019551$
$= 17922t + 1330.5$

$t = 97.5°F$

Cooling duty $[mc_p\Delta t]$

$\Delta h_{cooling} = (30)(8.3296)(1)(97.5-68)$
$= 7371.7$ Btu/min

$Q = mc_p\Delta t$

$m = 7371.7/[(1)(68-59)]$

$= 819.1$ lb/min $= 98.3$ gpm

7-14

$Q = mc_p\Delta t$

$Q_I = mc_p(250-t) = mc_p(170-40)$

$t = 120°C$

$Q_I = U_I A_I \Delta t_{lm} = 10\frac{kg}{s}(2000)\frac{J}{kg°C}(130)$
$= mc_p\Delta t$

$Q_I = 2.6\times10^6$ J/s $= U_I A_I \Delta t_{lm}$

$\Delta t = 80°C$

$A_I = Q/U_I\Delta t_{lm} = 2.6\times10^6/[500(80)]$

$A_I = 65.\ m^2$

$Q_{II} = 10\frac{kg}{s}\ 2000\frac{J}{kg°C}(250-170) = 1.6\times10^6\frac{J}{s}$

$A_{II} = 1.6\times10^6/\left[850\left(\frac{90-10}{\ln 90/10}\right)\right]$

$A_{II} = 51.7 m^2$

7-14 (con't)

Steam:

$Q_H = 1.6\times10^6\frac{J}{sec} = 1.6\times10^{-3}\frac{GJ}{s}$

$h_{fg} = 938.7\frac{Btu}{lb} = 2183.7\frac{J}{g}$

cost: $(\$3/GJ)(1.6\times10^{-3}\frac{GJ}{s})(3600)(24)(365)$
$= \$151373$/year

hxchr cost: $200(51.7+65) = \$23,340$

Cap cost $= 4(23,340) = \$93,360$

Annual cost $= 23,340 + 151373$
$= \$116700$/yr

7-15

Steam data

	1000 psia	3 psia
t, °F	544.75	141.4
h_f, Btu/lb		109.39
h_g, Btu/lb	1192.4	1122.5
h_{fg}, Btu/lb		1013.1
S_f, Btu/lb°R		0.20089
S_g Btu/lb°R	1.3903	1.8861
S_{fg} Btu/lb°R		1.6852

Turbine (constant entropy)

$1.39003 = 0.20089 + X(1.6852)$

$X = .705798$ quality of exit steam

$\Delta h = 1192 - [109.39 + X(1013.1)] = 367.97\frac{BTU}{lb}$

usable work: $367.97(.8)(.9)(2.928\times10^{-7}\frac{MWATT\text{-}hr}{Btu})$
$= (7.793\ MW\text{-}hr/lb)\times10^{-8}$

steam rate

$= 1 MW/7.793\times10^{-8}$

$= 1.2832\times10^7$ lb/hr steam

7-16

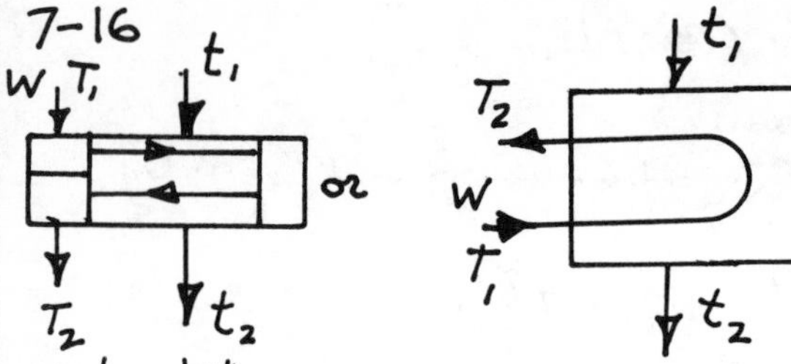

"straight" "hairpin"

Two types of 1-2 exchangers will be considered.

$W = 200{,}000$ lb/hr

$c_p = 1$ $U = 1000$

$\varepsilon = 0.00001$ (tubing is smooth)

$T_2 = 90$ $T_1 = 100$ $SG = 0.994$ (@95°F)

$t_2 = 90$ $t_1 = 75$ $\mu = .7194$ cp (@95°F)

$$X = \frac{T_2 - T_1}{t_1 - T_1} = 0.4$$

$$Z = \frac{t_1 - t_2}{T_2 - T_1} = 1.5$$

$Y = 0.8$ from Ten Broecke Chart

$$\Delta t_{lm} = \frac{(T_1 - t_2) - (T_2 - t_1)}{\ln \frac{T_1 - t_2}{T_2 - t_1}} = 12.33\ °F$$

$$q = W c_p [T_1 - T_2] = 2{,}000{,}000\ \frac{Btu}{hr}$$

$$A_{tubes} = \frac{q}{Y U \Delta t_{lm}} = 202.7\ ft^2$$

To calculate tube size the following input data for 18BWG tubing is needed
d is ID inches, A is external area ft²/ft, C is a factor used to calculate velocity

SIZE	d, Inches	A ft²/ft	C
3/8	0.277	0.0982	94
1/2	0.402	0.1309	198
5/8	0.527	0.1636	340
3/4	0.652	0.1963	521
7/8	0.777	0.2291	740
1	0.902	0.2618	997
1¼	1.152	0.3272	1626

For straight tubes, ½ of the tubes carry the total flow for each pass. Total length is 32 feet.

$$\therefore \#tubes = \frac{A_{tubes}}{(16)(2)\,A}$$

$$w = \frac{2W}{\#tubes} \qquad \frac{lb/hr}{tube}$$

$$v = \frac{w}{c\,(SG)} \qquad \frac{ft}{sec}$$

$$N_{Re} = 6.31 \frac{w}{d\mu}$$

Sacham equation

$$f = \left\{-2\log\left[\frac{\varepsilon/D}{3.7} - \frac{5.02}{N_{Re}} \log\left(\frac{\varepsilon/D}{3.7} + \frac{14.5}{N_{Re}}\right)\right]\right\}^{-2}$$

$$\Delta P = f\,\frac{32(12)}{D}\,v^2\,\frac{(SG)\,62.4}{2g\,144} \quad psi$$

SIZE	#tube	w	v	N_{Re}	f	ΔP
3/8	65	6200	66.4	196326	0.0161	657
1/2	48	8265	42.0	180326	0.0163	183
5/8	39	10329	30.6	171917	0.0164	75
3/4	32	12394	23.9	166732	0.0164	37
7/8	28	14465	19.7	163286	0.0165	21
1	24	16529	16.7	160734	0.0165	13
1¼	19	20659	12.8	157292	0.0166	6

Since the correct sized tube number is odd, round up to 20 each 1¼ 18BWG tubes
The resulting heat exchanger will have more area and less pressure drop than calculated. (a) 20 tubes
(b) 1¼ inch 18BWG each

7-16 (con't)

if hairpin tubes are used it is assumed that the length of run from tube sheet to tube sheet is 16 ft

$$\#\text{tubes} = \frac{A_{tubes}}{16A}$$

Since full flow in tubes

$$w = \frac{W}{\#\text{tubes}}$$

$$\Delta P = f\,\frac{16(12)}{D}\cdot\frac{v^2}{2g}\cdot\frac{(SG)62.4}{144}$$

SIZE	#tubes	w	v	N_{Re}	ΔP
3/8	129	1550	16.6	49081	27.1
1/2	97	2066	10.5	45082	7.6
5/8	77	2582	7.6	42979	3.1

7-17

let

$\sigma = .1713$ $\varepsilon = 0.8$ $h = 5$

$T_{SKY} = 460 - 40 = 420\,°R$

$T_{AIR} = 460 + 80 = 540$

heat balance

Roof is losing heat to sky by radiation, gaining heat by convection

$$\sigma\varepsilon\left[\left(\frac{T_{roof}}{100}\right)^4 - \left(\frac{T_{SKY}}{100}\right)^4\right] - h\left[T_{AIR} - T_{roof}\right] = 0$$

reduces to:

$$T_{roof}^4 + \frac{100^4 h}{\sigma\varepsilon}T_{roof} - h\frac{100^4}{\sigma\varepsilon}T_{air} - T_{SKY}^4 = 0$$

find at least 1 root between

7-17 (con't)

T_{air} & T_{SKY} : $T_{SKY} < T_{roof} < T_{air}$

using the "cosecant method of roots" only one root was found.

$T_{roof} = 527.3\,°R$ or $67.3\,°F$

7-18

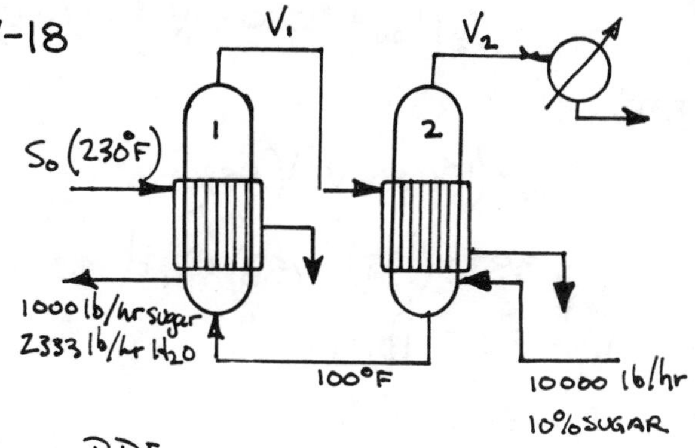

Ignore BPE

at 100°F VAP PRESS = 0.942 psi

Water balance

$$V_1 + V_2 + 2333 = 9000$$

$$V_1 + V_2 = 6667$$

assume $q_1 = q_2$

$$A_1U_1\Delta T_1 = A_1U_2\Delta T_2$$

$$400\Delta T_1 = 300\Delta T_2$$

$$\Delta T_1 = 230 - T_1$$

$$\Delta T_2 = T_1 - 100$$

$$t_1 = 174.3\,°F$$

Enthalpy balance with $t_1 = 174.3$

$t_2 = 100$

7-18 (con't)

EFFECT 2:

$$V_1 H_{174.3} + 10000(.95)(70) = V_2 H_{100} + (10000 - V_2)(.95)(100) + V_1 h_{174.3}$$

or

$$V_1 \lambda_{174.3} + 10000(.95)(70) = V_2 H_{100} + (10000 - V_2).95(100)$$

EFFECT 1:

$$S_0 \lambda_{230} + (10000 - V_2)(.95)(100) = 3333(.95)(174.3) + V_1 H_{174.3}$$

$V_1 = 3219$ lb/hr

$V_2 = 3448$ lb/hr

$S_0 = 3100$ lb/hr

$t_1 = 174.3^\circ \quad t_2 = 100^\circ$

$$Q_1 = S_0 \lambda = 3100(1157) = 3.586 \times 10^6 \text{ BTU/hr}$$

$$Q_1 = U_1 A_1 \Delta t = 400(A)(230 - 174.3)$$

$A_1 = 161 \text{ ft}^2$

$A_2 = 161 \text{ ft}^2$

7-19

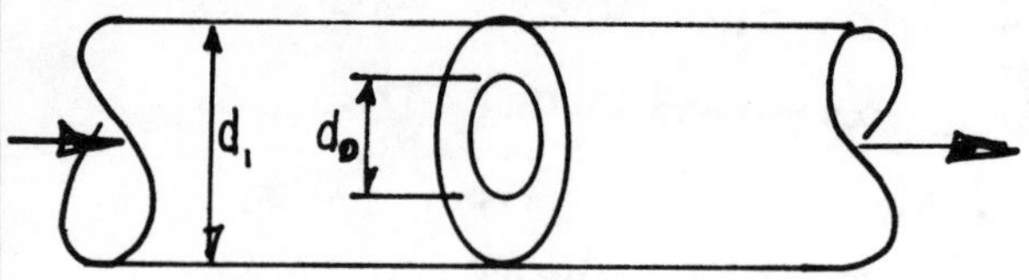

$100'' H_2O = 3.61$ psi

$$\frac{\Delta P}{P} = \frac{3.61}{50 + 14.7} = .0558$$

$$q_m = (412)\frac{Y d_0^2 c}{S_g}\sqrt{\Delta P \rho_1}$$

$$d_0^2 = \frac{1000(1)}{412\, Yc\sqrt{3.61(.07528)}} = \frac{4.65}{Yc}$$

$$[d_0]_{CALCULATED} = 2.1577 / \sqrt{Yc}$$

$$Re = \frac{q_H S_g}{d \mu} 482 = 264907$$

β guess	d_0	c	Y	d_0 calc
.3	1.8195	.6	.98	2.813
.4	2.426	.61	.975	2.7978
.5	3.0325	.625	.975	2.764
.457	2.77	.62	.975	2.775

$d_0 = 2.77$ inches

7-20

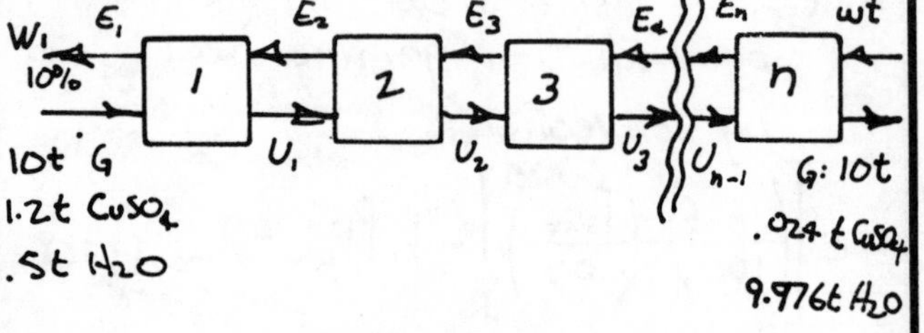

Assume: 1. Overflow & underflow lb/hr constant for stages 2 → N

2. Concentration exiting stage i = overflow exiting i

Overall H_2O bal.

$$0.5 + W = 19.976 + W_1 ; \quad W = 19.476 + W_1$$

7-20 (con't)

Recovery: $1.2 - .024 = 1.176$t $CuSO_4$

∴ $1.176 = 0.1(W - 19.476 + 1.176)$

$W = 30.06$t H_2O input

∴ for stages 2 → n

$U = 20$t /hr

$E = 30.06$t /hr

$y_i = x_i$

stage 1

$E_1 + U_1 = U_0 + E_2$

$CuSO_4$:

$1.2 + y_2 E_2 = x_1 U_1 + y_1 E_1$

$y_2 = [.1(20 + 11.76) - 1.2]/30.06$

$y_2 = 0.0657$ (6.57%)

stage 2 → n

$E_i y_i + U_i x_i = E_{i+1} y_{i+1} + U_{i-1} x_{i-1}$

∴ $y_i = \left(-\frac{U}{E}\right) x_i + y_{i+1} + \frac{U}{E} x_{i-1}$

but $y_i = x_i$

$$y_i = \frac{y_{i+1} + \left(\frac{U}{E}\right) x_{i-1}}{1 + \frac{U}{E}}$$

for stagewise calculation:

$$y_{i+1} = y_i\left(1 + \frac{U}{E}\right) - \frac{U}{E} x_{i-1}$$

$$= y_i\left(1 + \frac{U}{E}\right) - \left(\frac{U}{E}\right) y_{i-1}$$

$U/E = 20/30.06 = 0.66534$

$y_{i+1} = y_i(1.66534) - 0.66534\, y_{i-1}$

7-20 (con't)

i	y_{i+1}
1	.0657
2	.0429
3	.0278
4	.0177
5	.0110
6	.0065
7	.0035
8	.0015
9	.0002

9 stages